Management of Tropical Freshwater Fish Ponds

by

D.K. Chatterjee

S. Adhikari

2018

Daya Publishing House®

A Division of

Astral International Pvt. Ltd.

New Delhi - 110 002

First Published, 2009
Reprinted, 2018

ISBN: 978-81-7035-607-3 (Hardbound)

Published by : **Daya Publishing House®**
A Division of
Astral International Pvt. Ltd.
– ISO 9001:2015 Certified Company –
4736/23, Ansari Road, Darya Ganj
New Delhi-110 002
Ph. 011-43549197, 23278134
E-mail: info@astralint.com
Website: www.astralint.com

Foreword

Tropical freshwater fish pond aquaculture has been drawing increasing attention as a reliable technology for raising food fish. Both from viewpoints of producing nutritious tasty food and as a profitable venture particularly in rural areas, the enterprise is gaining increased attention.

With availability of seed of desirable fish species as a result of spread of practice of induced breeding of fishes, planned fish culture in tropical ponds is increasingly being practiced.

In these circumstances, it is essential to know the fertility status of the ponds *i.e.* the water and soil quality of the water bodies for their use in fish culture. This guides the planning of the culture process for obtaining good fish crops and has led to wide adoption and use of scientific methods of fish production from tropical ponds.

The authors of the book have long experience of research work in water and soil chemistry as applicable to scientific management of fish ponds in tropical conditions both farm ponds as well as privately owned ponds.

Exhaustive but easy to follow methodologies to be adopted for profitable fish culture have been described in the book. The authors drawing on their practical experience, approach their subject in details at all stages from selection, planning and management of ponds in tropical conditions, in easy language which makes the book an ideal manual on the subject.

Though exhaustive, the book presents required information in easy to understand language for practice of successful fish culture and as a dependable guide in planning and execution of fish culture work. It would serve the needs of the student, teacher, extension worker or developmental agencies and as such has wide utility.

The long experience of work of the authors has made writing of this valuable book possible. Suggestions, recommendations etc, however the reader may take the advice rendered herein, needs to be treated as dependable advise.

The book fills a crying need for a guide in the field of water and soil chemistry for the practice of successful pond fish culture and what better than being based on long experiences of dedicated scientists acquired in the course of work for the premier freshwater research Institute of the country, the CIFA at Kausalyaganga, Orissa.

May the book have wide readership and render service in the cause of spread of freshwater aquaculture.

R.D. Chakrabarty

Retired Head of Division

Freshwater Fish Culture,

CIFA, Kausalyaganga, Orissa, India

Preface

The need for maintaining a proper nutrients status in the pond environment for sustained fish production has been well appreciated in the Indian Freshwater Aquaculture practices. Being an agrarian economy the country's aquaculture practices have also been largely organic based, utilizing the agro-livestock-byproducts. The culture systems have also been suitably modified to utilize as well treats the wastewater, particularly the domestic sewage that has become a well-known system of freshwater aquaculture.

Intensification of aquaculture practices in the recent years has posed question with regard to maintenance of optimum nutrient status in the culture system as well as the impact of the practices on the surrounding environment. The new approach to nutrient management in the ponds gives due attention to these aspects of organic decomposition, residual accumulations, possibilities of microbial processing, fertilization strategies *to suit* different kinds of the ponds; biofertilization, biofiltration, etc. to achieve increased productivity, sustainability and environment upkeep. Aquatic environmental management is crucial at a stage when efforts are being made to increase the mean national production to a tune of at least 5 tonnes/ha/yr and at the same time, diverse nutrient resources are being available for recycling and utilization in fish culture.

The present book is planned for providing the readers/fish culturists an exposure to the principles and practices of fertilization in freshwater aquaculture, recent trends in nutrient management, soil and water quality management of fishponds, role of microbes in nutrients recycling in fishponds and to the tools that could improve aquaculture productivity. It is hoped that the book could help the fish culturists with theoretical aspects as well as practical tools in the management of fishponds for optimum and sustainable production.

D.K. Chatterjee

S. Adhikari

Contents

List of Figures

Chapter 1
Introduction

Water quality variables are primary factors affect production of freshwater fish pond. In depth informations/ publications on water quality dynamics in ponds and methods of management are available. However, a simple and concise discussion of the major aspects of water quality and soil conditions and their management is needed.

The major water quality variables, salinity, pH, Oxygen, Plankton, Nutrients and Toxic metabolites including Physico and Chemical Characteristics of pond soil are covered. These variables are discussed in relation to use of chemical fertilizers and manure to increase fish production under tropical conditions. A brief mention of effects of aeration and feeding on water quality have also been included.

Although feeding and aeration are not extensively used, these techniques will no doubt gain acceptance in future. In addition, pond environment conditions due to aquaculture practices have been presented.

Chapter 2

Photosynthesis, Respiration and Aquaculture

In ponds, phytoplankton are the primary source of Organic matter, that ultimately finds its way into fish. Plants have the ability to use carbondioxide (CO_2), Water (H_2O), mineral nutrients and sunlight to produce organic matter as simple sugar ($C_6H_{12}O_6$), and produce oxygen (O_2) as a by-product. The process, termed as photosynthesis, inorganic carbon present in carbondioxide is chemically reduced to organic carbon as sugar and light energy (sunlight) is transformed to chemical energy of sugar. The summary reaction of photosynthesis is:

$$\text{Light Energy} + 6CO_2 + 6H_2O \rightarrow C_6H_{12}O_6 + 6O_2$$

The simple sugar molecules from photosynthesis by green plants represent nearly all the energy available to living things. Both plants and animals depend upon photosynthetically produced energy. The simple sugar molecules also are the building blocks for more complex organic compounds. Plants make complex carbohydrates (starch, cellulose etc.) proteins, fats, vitamins and other compounds from the sugars from photosynthesis. Plants form their structure from these compounds and use photosynthetically derived sugar as source of energy.

Animals cannot produce organic matter. They must feed directly on plants or on other animals that have fed on plants. Therefore, both energy and structural materials needed by animals come originally from plants.

Respiration is a second basic process in aquaculture. In respiration organic matter combine with oxygen (Oxidised) to release water, carbondioxide and energy. Plant and animal cells have the ability to capture some of the energy released through oxidation and use it to do biological work. From an ecological stand point, respiration is the reverse process of photosynthesis :

$$C_6H_{12}O_6 + 6O_2 \rightarrow 6CO_2 + 6H_2O + \text{heat energy}$$

In photosynthesis, carbondioxide is reduced to organic carbon with the capture of energy and the release of oxygen, while in respiration organic carbon is oxidised to carbondioxide with the release of energy and the uptake of oxygen. Of course from a bio-chemical stand-point photosynthesis and respiration are quite distinct processes, but for ecological purposes photosynthesis and respiration may be thought of a reversible reactions. When photosynthesis is progressing faster than respiration, oxygen will accumulate and carbon dioxide will decline and is the usual situation in a pond during day light. At night, photosynthesis stops/ceases but respiration occurs day and night. Thus, at night, oxygen declines and carbondioxide increases.

The food chain or food web in a fish pond (Figure 1) initiates with plants. In ponds, phytoplankton are the most desirable plants. These organisms are microscopic algae that swim or remain suspended in the water. Algae often are green in colour, but some may be blue-green, yellow red, black or brown. When pond water contain a "Phytoplankton bloom" or more generally, a "Plankton bloom,"the phytoplankton may be fed upon by microscopic animals called zooplankton. Collectively phytoplankton and zooplankton are called simply plankton. The plankton may die and fragmented to form dead organic matter (detritus) which is food for bacteria and various animals. Algae can grow on the pond bottom where there is sufficient light for photosynthesis. Detritus settles to the pond bottom, thus enriches the soil with organic matter. The pond bottom supports a community of bacteria, algae and small animals which is/are called benthos. Aquatic insects which feed on plankton, benthos or detritus also abound in ponds. In the shallow areas of ponds with clear waters, larger aquatic plants (macrophytes) may grow. Aquatic macrophytes also may grow while floating on the pond surface. Because few fish or other animals feed directly on macrophytes, and because the large structure of macrophytes obstructs the water. Macrophytes are undesirable in most fish ponds. As far the fish depending on the species, they eat plankton, benthos, aquatic insects, detritus, or some combination of

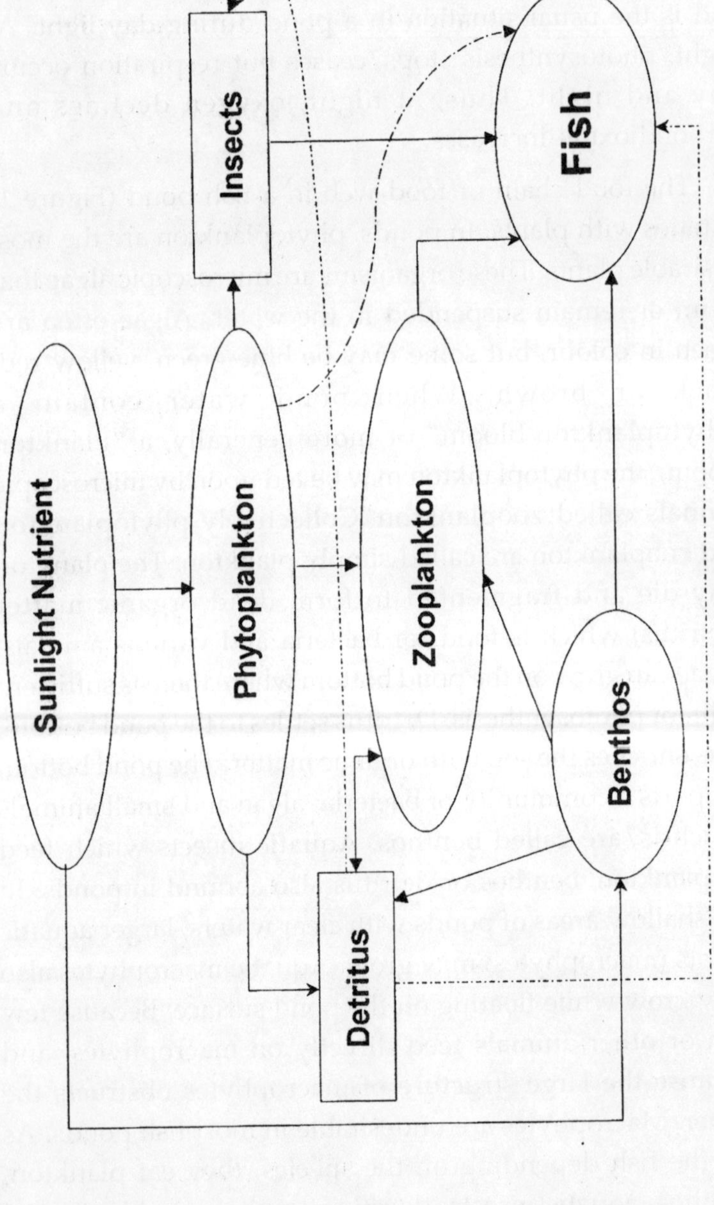

Figure 1: Food Chain in Fish Pond Ecosystem

these food organisms. Some species of fish even eat macrophytes and can be used for controlling macrophytes.

In order to increase fish production in ponds, it is necessary to increase the amount of food for fish. This can be done by improving conditions for production of phytoplankton, which in turn will increase the production of other fish food organisms. Usually, it is only necessary to add to ponds certain inorganic nutrients in the form of manure or chemical fertilizers to increase, the growth of phytoplankton. Of course in fish culture, it is possible to add artificial feed for fish and short circuit the food-chain. Neverthless, feeding does not change the dependence of fish on plants, for the fish feed is made from a plant based eco-system.

It will be seen later that phytoplankton is extremely important in the dynamics of oxygen concentrations in ponds. As phytoplankton growth is enhanced to foster greater fish production, wide changes in oxygen concentrations in water occurs between day and night. Excessive phytoplankton blooms may lead to oxygen depletion and fish stress or mortality. Water quality in ponds is to a large degree dominated by phytoplankton abundance and the balance between photosynthesis and respiration.

Chapter 3
Substances in Water

It is essential for the aquaculturists to have an understanding of the substances that are dissolved or suspended in pond water. The concentrations of certain substances are vital to the well being of a fish pond. In aquaculture, water quality refers to suitability of the water for fish growth. There are many water quality variables, but fortunately only a few are vital in pond fish culture.

3.1 Inorganic Substances

Dissolved inorganic substances in water include almost every element in the earth's crust. However, the most important inorganic elements, ions or complexes are those needed for plant and animal nutrition and metabolism and those that are toxic.

A large number of inorganic elements are required for plant growth. Most species require at least the following:

Carbon, hydrogen, oxygen, nitrogen, sulphur, phosphorus, chloride, boron, molybdenum, calcium, magnesium, sodium, potassium, zinc, copper, iron and manganese. Diatoms also require silicon. Aquatic plants make oxygen in photosynthesis, and they obtain hydrogen from water. Carbon dioxide enters water from the air and from decomposition of organic matter. The other elements enter ponds from the water supply, from dissolution of minerals in the pond bottom, or in additions of manure or fertilizer. Of course, some algae and bacteria are able to fix nitrogen. That is they can take molecular nitrogen (N_2), which enters water from the air, and convert this nitrogen to nitrogen in plant biomass.

Nitrogen and phosphorus are most likely to limit phytoplankton growth. Typical concentrations of plant nutrients in pond water and in phytoplankton biomass are presented. There is less nitrogen and phosphorus in pond water relative to phytoplankton needs than for other elements. Hence, fertilizers and manures are added to ponds to supplement the natural shortage of nitrogen and phosphorus.

After nitrogen and phosphorus, carbon is the next most common element to limit productivity in fish ponds. Of course, carbon supplies in ponds can be augmented by application of manure. As will be seen later, carbon is contained in the alkalinity system and lime applications to increase alkalinity improve carbon availability. The natural availability of carbon is particularly low in acidic ponds.

Thus, the three major substances used to promote phytoplankton and increase fish production are manures, chemical fertilizers, and lime. There are only a few reported instances where nutrients other than carbon, nitrogen and phosphorus limit phytoplankton in ponds.

The forms of the various inorganic substances and their desired ranges are provided in Table 1.

Table 1

Element	*Form in Water*	*Desired Concentration*
Oxygen	Molecular Oxygen (O_2)	5 to 15 mg/l
Hydrogen	H^+	PH 6-9
Nitrogen	Molecular Nitrogen N2	Saturation or less
	Ammonium (NH_4^+)	0.2–2.0 mg/l
	Ammonia (NH_3)	< 0.1 mg/l
	Nitrate (NO_3)	0.2–1.0 mg/l
	Nitrite (NO_2)	< 0.2 mg/l
Sulphur	Sulphate (SO_4^{2-})	5-100 mg/l
Carbon	Carbondioxide (CO_2)	1-10 mg/l
Calcium	Calcium ion (Ca^{2+})	5-100 mg/l
Magnesium	Magnesium ion (Mg^{2+})	5-100 mg/l
Sodium	Sodium (Na+)	2-100 mg/l
Potassium	Potassium ion (K+)	1-10 mg/l
Bicarbonate	Bicarbonate ion (HCO_3)	20-300 mg/l
Carbonate	Carbonate ion (CO_3^{2-})	0-20 mg/l
Chloride	Chloride ion (Cl)	1-100 mg/l
Phosphorus	Phosphate ion (HPO_4^{2-}, H_2PO_4)	0.005-0.2 mg/l
Silicon	Silicate (H_2SiO_3, $HsiO_3$)	2-20 mg/l
Iron[1]	Ferrous iron (Fe^{2+})	0 mg/l
	Ferric iron (Fe^{3+})	Trace
	Total Iron	0.05-0.5 mg/l
Manganese[1]	Manganese ion (Mn^{2+})	0 mg/l
	Manganese dioxide (MnO_2)	Trace
	Total Manganese	0.05-0.2 mg/l

Contd...

Table 1–Contd...

Element	*Form in Water*	*Desired Concentration*
Zinc[1]	Zinc ion (Zn^{2+})	< 0.01 mg/l
	Total Zinc	0.01 to 0.05 mg/l
Copper[1]	Copper ion (Cu^{2+})	< 0.005 mg/l
	Total copper	0.005 to 0.1 mg/l
Boron	Borate (H_3BO+, H_2BO_3)	0.05–1 mg/l
Molybdenum	Molybdate (MO_4)	Trace

Seven ions (Na^+, K^+, Ca^{2+}, Mg^{2+}, Ca, SO_4^{2-}, HCO_3^-) normally contribute 75 per cent or more of the weight of the dissolved ions in water. However, many of the other ions are extremely important biologically inspite of their relatively low concentrations.

The desirable ranges for these substances are poorly understood. The values listed as the desired concentrations are actually the usual concentrations of these six trace metals in surface waters of ponds.

In addition to dissolved inorganic substances, pond water may contain suspended inorganic soil particles. These particles usually enter ponds in the water supply (run off is often turbid), or they are suspended in water by heavy wave action. The larger particles will settle to the pond bottom, but some of the smaller particles may remain suspended for long periods and cause turbidity. It is undesirable for fish culture when turbidity by soil particles restricts visibility of light into the water to less than 20-25 cm.

3.2 Organic Substances

A wide range of organic substances occur in pond waters. Dissolved compounds include sugars, starches, amino-acids, polypeptides, proteins, fatty acids, tannins, humic acids, vitamins, etc. large particles of decaying organic matter called detritus also are plentiful. Of course, the plankton and bacteria contribute to the organic load in water. It is not covenient to analyse the specific organic compounds. Usually the total weight of organic matter or the total weight of particulate organic matter in water is determined. Desirable ranges for organic matter concentrations are not known, but pond water usually contains 10 to 200 mg of organic water.

Organic substances in water, particularly plankton, cause turbidity. Contrary to suspended clay particles, turbidity by plankton is desirable. Ponds are most productive when turbidity by plankton restricts visibility in water to 20 to 40 cm. At this level of plankton abundance, there is usually adequate food and oxygen for fish and light does not penetrate to the pond bottom to encourage growth of rooted aquatic macrophytes. Unfortunately, floating aquatic macrophytis such as water hyacinth (*Eichhornia crassipes*), water lettuce (*Pistia stratiodes*), duckweed (*Lemna spp.*) etc. cannot be controlled by turbidity.

Chapter 4
Gross Variables

For analytical convenience and practical utility, a number of gross characteristics of water are frequently determined. These concentrations of variables are meaningful to determine the capacity of a particular water for fish production.

4.1 Salinity

The total concentrations of all dissolved ions are expressed as the salinity of water. In fresh water, salinity is usually expressed in milligrams per litre. In humid areas, inland waters usually contain 50 to 1000 mg/l, salinity. Water with more than 500 mg/l salinity usually is not suitable for domestic purposes, and 1,000 mg/l salinity will impart a salty taste. In arid regions and even during the dry season in certain humid areas, inland waters can become quite saline. Most fresh water fish can do well in waters with salinities up to 2,000 mg/l, some species tolerate even much higher salinities.

4.2 Total Dissolved Solids

The total concentration of all dissolved substances in water is the total dissolved solids. Usually concentrations of salinity and total dissolved solids are similar.

4.3 Total Alkalinity

The total concentration of bases in water expressed in milligrams per litre of equivalent calcium carbonate is the total alkalinity. There are several bases in water : OH^-, NH_3, $H_2BO_3^-$, HPO_4^{2-}, $HSiO_3$, HCO_3^- and CO_3^{2-}, but in most pond waters HCO_3^- and CO_3^{2-} greatly exceed than other bases in concentration.

The total alkalinity in water is divided primarily from dissolution of limestone in soils, primary by soil characteristics. For example, ponds in areas with sandy soils often have total alkalinities below 20 mg/l, while ponds in areas in calcareous soils may have total alkalinities above 100 mg/l. Other factors being equal, total alkalinity will be higher in ponds in arid regions than those of humid areas.

As a general rule, the natural fertility of pond water increases with increasing total alkalinity up to at least 150 mg/l. However it is generally accepted that ponds with total alkalinities above 20 mg/l can be productive for fish. Alkalinities below 20 mg/l require liming. The ideal range for total alkalinity is considered to be in the range of 50 to 100 mg/l. While more than 100 mg/l does not affects the productivity.

4.3.1 Total Hardness

The total concentration of all divalent cations in water expressed in terms of milligrams per litre of calcium carbonate is the total hardness. Calcium and magnesium are the dominant divalent cations in most pond waters. As a general rule, hardness, like alkalinity is derived from the dissolution of lime stone. If a quantity of limestone dissolves, it gives equal amounts of hardness and alkalinity. Therefore in most waters, total hardness and total alkalinity concentrations remain approximately equal. There are, however, some notable exceptions. In arid regions and in regions where soils are quite acidic, waters typically have greater hardness than alkalinity. In some coastal areas, well waters may have much higher alkalinities than hardness. Photosynthesis cause high pH in ponds filled with such water. The desirable range for total hardness remain same as for total alkalinity.

4.3.2 Acidity

Carbon dioxide is acidic in water, but normally it cannot depress pH below-4.5. Waters that have a lower pH contain a strong mineral acid-usually sulphuric acid. Such waters do not contain alkalinity, and are unsuitable for fish. The mineral acidity of water is a measure of the total acids in water, expressed in terms of milligrams per litre of equivalent calcium carbonate. In fact mineral acidity can be considered as negative alkalinity, because it represents the amount of calcium carbonate that would have to be added to water to raise the pH to the point that addition of more calcium carbonate would cause alkalinity.

4.4 Biochemical Oxygen Demand

This is a measure of the rate of oxygen consumption by the plankton and bacteria in a sample of pond water. A sample of raw water or diluted water is incubated in the dark for 5 days at 20°C. The loss of Dissolved Oxygen (DO) from the water during the incubation period is the Biochemical Oxygen Demand (BOD). Fish ponds typically have BOD values 5 to 20 mg/l. The higher the BOD, the greater the degree of enrichment of pond water with organic matter. Although BOD often has been measured in pond waters, the desirable range is not well defined. Oxygen depletion is a danger when the BOD exceeds 10 to 20 mg/l.

4.4.1 Chemical Oxygen Demand

This variable is measured by converting all of the organic matter in a water sample to carbon dioxide and water by oxidation with potassium dichromate and sulphuric acid. The amount of potassium dichromate consumed in the oxidation is measured and the oxygen equivalent of the dichromate is the chemical oxygen demand (COD). Thus, the same amount of oxygen would be necessary to oxidise completely the same amount of organic matter represented by C.O.D. The COD is an index of the organic enrichment of a pond water. Fish ponds usually have COD values of 10 to 200 mg/l. The best range is probably 40 to 80 mg/l.

4.5 Secchi Disc Visibility/Turbidity

The Secchi disc is a 20 cm diameter disc painted with alternate black and white quadrants. It is weighted under the bottom and attached at the centre of its upper surface with a calibrated line. The depth at which the disc just disappears from view is secchi disc visibility. Obviously care must be taken to standardise the procedure for reading the secchi disc. In many waters, there is a close correlation between Secchi disc visibility and plankton abundance. As plankton density increases, visibility decreases. However, waters containing much turbidity from suspended clay particles or detritus, the secchi disc visibility will not be suggestive of phytoplankton abundance. A secchidisc visibility of 20 to 40 cm is indicative of a productive fish pond, provided the turbidity is primarily from plankton.

4.5.1 Chlorophyll

It is possible to measure chlorophyll concentration and use it as an index of the abundance of phytoplankton. In general, as the chlorophyll concentration increases the phytoplankton abundance also increases. Productive fish ponds often have a chlorophyll concentrations of 50 to 200 µg/l (0.05 to 0.2 mg/l).

4.6 Primary Productivity

A fairly accurate assessment of phytoplankton productivity in fish ponds can be made by light dark bottle productivity estimates. The general procedure involves filling three BOD bottles with pond water. One bottle called the initial bottle is used to determine the initial DO concentrations. The dark bottle is wrapped with material to exclude light. The light and dark bottles are incubated in the pond for a specific period of time, removed and the DO concentration is determined. Oxygen concentration in the bottles may be used to estimate respiration and photosynthesis.

(*i*) Initial Bottle-Dark Bottle = Plankton Respirations

(*ii*) Light Bottle-Dark Bottle = Gross Oxygen Production

In most application, only gross photosynthesis is measured and the initial bottle is unnecessary. Of course, determination of respiration a short BOD under actual pond conditions is obtained. If desired, oxygen data can be converted to carbon values, because every molecule of O_2 produced in photosynthesis is a molecule of CO_2 is converted to organic matter and vice versa. The molecular weight of carbon is 12 (there is one C in CO_2) and that of oxygen is 32. Hence, multiplication of gross oxygen production by 0.375 gives gross photosynthesis in mg C/1. Usually gross photosynthesis is reported on a square meter basis. Typical values for fish ponds range from 2 to 6 mg C/m^2/day. In general fish production increases with increasing gross photosynthesis.

4.7 pH

The term pH refers to the hydrogen ion concentration in water, more generally, pH refers to the acidic or basis nature of water. For practical purposes, water with a pH of 7 is considered neither acidic nor basic, it is said to be neutral. When the pH is below 7, the water is said to be acidic. Water with a pH above 7 is considered basic. The pH scale extends from 0 to 14; the greater the difference in the pH and 7, the more acidic or basic is the water.

4.7.1 Factors Affecting pH

The pH of normal fresh water is between 6 and 8. Phytoplankton bloom develop in ponds. Phytoplankton obtains CO_2 for use in Photosynthesis from the CO_2 – HCO_3 – CO_3^{-2} equilibrium as follows:

$$2\ HCO_3 = CO_2 + CO_3^{2-} + H_2O$$

As CO_2 is removed, the reaction progresses to the right of the equation and CO_3^{2-} accumulates.

Because phytoplankton uses CO_2 during daylight, the pH of water increases during the day. At night, no CO_2 is removed by phytoplankton, but all pond organisms release CO_2 from respiration. This CO_2 reacts with CO_3^{2-} and H_2O to form HCO^-_3, the HCO^-_3 dissociates to release H^+ and the pH declines. The daily fluctuation in pH is not always great, but wide pH fluctuation can result when phytoplankton is abundant.

Removal of CO_2 will make the water less acidic with the increase in pH, while addition of CO_2 will make the water more acidic with reduction in pH, Carbon dioxide is an acid because it reacts with water to yield H^+ ions.

$$CO_2+H_2O=HCO_3^-+H^+$$

The pH of pure water (distilled water) saturated with CO_2 is around 5.6.

4.7.2 Effects on Fish

The direct influence of pH on fish is generalized in Table 2.

Table 2

pH	*Effect*
4	Acid death point
4-5	No reproduction
5-6	Slow growth
6-9	Best growth
9-11	Slow growth
11	Alkaline death point

Where the pH of pond water is too low, lime may be applied to improve pH. Low pH is more common than high pH. This is fortunate, for there is no really reliable procedure for reducing high pH.

Usually, acidity problems in pond do not result from direct effects of low pH on growth, reproduction, or survival, but from the effects of low alkalinity and acidic muds on plankton and benthic production. The effects of course, are reflected in low production of fish.

4.8 Dissolved Oxygen

Dissolved oxygen is the most critical water quality variable in aquaculture. It is essential for the aquaculturists to understand thoroughly the factors affecting the concentration of DO in pond water and be aware of the influence of low DO concentrations on fish.

4.8.1 Solubility

The atmosphere contains 20.95 per cent oxygen at standard barometic pressure (760 mm of mercury), the pressure of oxygen in air is 159.2 mm (760 × 0.2095). The pressure of oxygen in air drives oxygen into water until the pressure of oxygen in water is equal to the pressure of oxygen in air. When the pressure of oxygen in water and air are equal, net movement of oxygen from air to water ceases and DO is said to be at equilibrium or saturation.

The solubilities of DO at saturation for standard barometic pressure and different temperatures are provided in Table 3 for freshwater. The concentration of DO at saturation declines marked by as water temperature increases. The concentration of DO at saturation also declines with increasing salinity, but this effect is not great over the salinity range for fresh water aquaculture. The concentration of DO at saturation decreases with decreasing barometic (atmospheric) pressure. Variation in barometric pressure at a given locality may be ignored, but variations in barometic pressure resulting from elevation must be taken into account when using data from Table 3.

Table 3: Solubility of Oxygen in Pure Water at Different Temperature

°C	mg/l	°C	mg/l	°C	mg/l
0	14.16	12	10.43	24	8.25
1	13.77	13	10.20	25	8.11
2	13.40	14	9.98	26	7.99
3	13.05	15	9.76	27	7.86
4	12.70	16	9.56	28	7.75
5	12.37	17	9.37	29	7.64
6	12.06	18	9.18	30	7.53
7	11.76	19	9.01	31	7.42
8	11.47	20	8.84	32	7.32
9	11.19	21	8.68	33	7.22
10	10.92	22	8.53	34	7.13
11	10.67	23	8.38	35	7.04

(For an atmosphere saturated with water vapour at a pressure of 760mm of Hg)

Where barometric pressure is unknown, the approximate change in pressure with increasing elevation is as follow: 0 to 600 m-4 per cent decrease in BP per 300 m, 600 to 1500 m-3 per cent decrease in BP per 300m, 1,500 to 3,000m-2.5 per cent decrease in BP per 300m. However the influence of elevation on the solubility of DO can be ignored except for rigorous calculations.

Pressure at a point in water is influenced by the depth of the point beneath the water surface. The weight (pressure) of water above the point is called the hydrostatic pressure, and the total pressure at the point is hydrostatic pressure plus barometric pressure. The solubility of DO at saturation at the point is a function of total pressure. Thus, the solubility of DO at saturation increases with increasing water depth. However, the effect of depth on the solubility of DO at saturation normally is ignored due to shallow depth of fish ponds.

Plants growing in pond water produce oxygen by photosynthesis, and during day light, plants may produce oxygen so fast that DO concentrations in water rise above saturation. Water containing more DO than expected for the existing barometric pressure and water temperature is said to be super saturated with DO.

Water also may contain less DO than expected at saturation for prevailing conditions. Respiration by organisms in ponds may cause DO levels to decline; DO typically declines below saturation at night.

A net movement of oxygen molecules from air to water occurs when water is below saturation with DO. At saturation with DO, the number of oxygen molecules leaving the water equals the number entering; there is no net movement of oxygen molecules. Net movement of oxygen molecules from water to air occurs when water is super saturated with DO. The larger the difference between the pressure of oxygen in water and air, the greater is the net exchange of oxygen molecules.

The degree of saturation of water with DO frequently is expressed as percentage saturation. The equation for estimating percentage saturation is:

$$\% \text{ Saturation} = \frac{\text{DO concentration in water}}{\text{DO concentration at saturation}} \times 100$$

Effects on Fish

The influence of DO concentrations on fish cultured in ponds is summarized in Table 4.

Table 4

DO Concentration	*Effect*
Less than 1 or 2 mg/l	Lethal if exposure lasts more than a few hours.
2.0-5.0 mg/l	Growth will be slow if exposure to low DO is continuous.
5 mg/l–Saturation	Best condition for good growth.
Above saturation	Can be harmful if saturated conditions exist throughout pond volume.

Concentrations of DO can fall so low that mortality of fish in ponds may occur. However adverse effects of low DO more often are expressed as reduced growth and greater susceptibility to disease. In ponds with chronically, low DO concentrations, efficient conversion of food to fish flesh will be reduced compared to ponds with normal DO concentrations.

4.8.2 Plankton and DO

Light passing through pond water is quenched rapidly, and the rate of quenching increases as the amount of particulate matter (turbidity) in the water increases. As a result photosynthesis occurs most rapidly in the surface layer of water, and DO concentrations decline with-depth. Plankton blooms reduce light penetration, and the amount of light available for photosynthesis at a given depth is proportional to the amount of plankton. In ponds with a lot of plankton, DO concentrations may fall to 0 mg/l at depths of 1.5 to 2 m. Because of this, it is best to use shallow ponds of 0.75 to 1.5 m for aquaculture where climatic condition and water supply permits. Of course, in areas with a long dry season, it may be impossible to maintain adequate water depth in shallow ponds. Concentration of DO exhibit a daily cycle. The lowest concentrations of DO occur about dawn. During day light, photosynthesis causes DO concentrations to increase, and maximum DO concentrations are reached in the afternoon. During the night, photosynthesis ceases but continuing use of oxygen by pond organisms cause DO concentrations to decline. The daily cycle in DO is most pronounced in ponds with heavy phytoplankton blooms. The influence of the daily cycle of DO on growth of aquaculture species is poorly understood, but most workers feel that good growth can be achieved as long as the DO concentration does not fall below 25 or 30 per cent of saturation during the night and does not remain at this low level for more than 1 or 2 hours.

Cloudy weather can influence DO concentrations. This results because cloudy weather reduces the rate of photosynthesis through light limitation, but it has little or no effect on respiration. The influence of cloudy weather is more pronounced in a pond with a heavy phytoplankton bloom than in a pond with less phytoplankton.

Increased manuring or fertilization rates in fish ponds also increases phytoplankton abundance in ponds. This permits higher fish production, but it also causes DO concentrations to fluctuate widely between day and night and with depth. At higher rates of manuring or fertilization, phytoplankton blooms will become so dense that fish growth will decline or fish will die because of low DO concentrations. The manuring or fertilization rates must be suitably adjusted so that there is both adequate plankton and DO for good fish production. Because of differences in composition of manures and differences in responses of individual ponds to manures and fertilizers, it is not possible to recommend a single manure or fertilizer application rate and schedule for all ponds. It is essential that manure or fertilizer applications is adjusted to fit pond conditions.

Phytoplankton in ponds may suddenly die and decompose and thereby cause depletion in DO in ponds. The DO concentrations did not return to normal until a new phytoplankton bloom was established. Most phytoplankton die-offs involve species of blue-green algae. During calm weather, blue-green algae often form scums at pond surfaces. Intense sunlight may result in sudden death of algae in this scum. Blue-green algae decompose rapidly due to high concentration of nitrogen in their tissues.

Mats of filamentous algae which develop on pond bottom may, under certain conditions, float to the surface of a pond and die. This phenomenon also can cause depletion of DO.

4.8.3 Bottom Sediment and DO

Although plankton abundance is the dominant factor in DO dynamics in fish ponds, the bottom sediments also consume DO. Bottom sediments, especially in old ponds, where large amounts of organically-enriched sediment have accumulated in pond bottom, may exert oxygen demands. There has been little information on DO consumption rates by pond soils, but there is evidence that respiration by the benthic community can easily remove 2 to 3 mg/l of DO from pond water in 24 hours.

4.8.4 Feedings and DO

It has already been stated/mentioned that phytoplankton abundance is controlled by nutrient supply and that DO concentrations are regulated to a large extent by phytoplankton abundance. Feed applied for fish results in pollution of pond waters by organic and inorganic metabolic wastes from fish. Unutilized feed also release nutrients into the water after decomposition. Consequently, phytoplankton abundance and problems with low DO increase as a function of increasing feeding rate. Data suggests that feeding rate above 40 to 50 kg/ha per day will result in unacceptably low DO. Higher feeding rates may be used in ponds if aeration is applied.

Feed conversion value is determined as the quantity of feed applied divided by the weight of fish harvested. A low feed conversion value indicates greater efficiency than a high value. With good management practices, feed conversion values of 1.5 to 2.0 may be achieved with many species of fish. One of the effects of overfeeding in fish ponds is to increase the feed conversion value. As feeding rate increases, DO concentrations during the night decline. Chronically low DO concentrations have an adverse effect on the apetite and metabolism of fish and feed conversion values tend to increase drastically if feeding rates are increased to a level where DO concentrations fall below 2 or 3 mg/l each night.

Chapter 5
Nitrogen and Phosphorus

Nitrogen and phosphorus are the two most important nutrients with respect to phytoplankton growth. Summary of the dynamics of these two plant nutrients is provided.

5.1 Nitrogen

The nitrogen cycle is given in Figure 2 Nitrogen may enter ponds from the atmosphere in molecular form (N_2), and some molecular nitrogen can be fixed as inorganic compounds by blue-green algae and bacteria. Rain falling in to ponds contains nitrate, and various forms of nitrogen may be added in fertilizer and organic nitrogen in manure or feed. In the pond, nitrogen undergoes many transformations through biological activity as discussed below. However, the dissolved inorganic nitrogen content

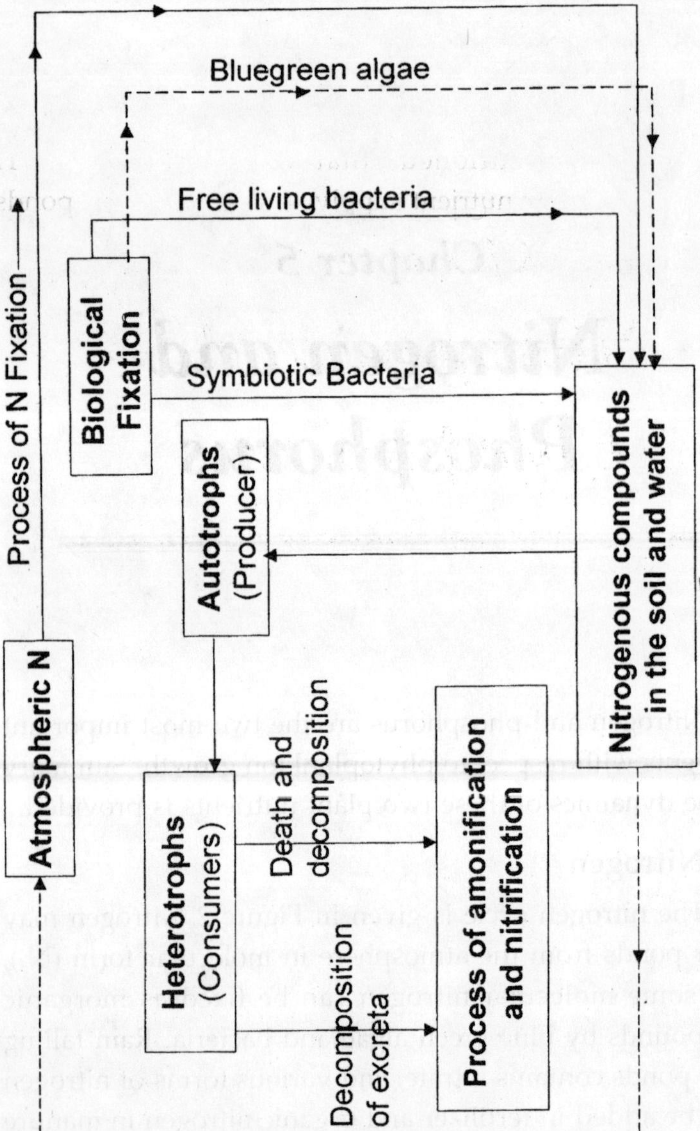

Figure 2: Nitogen Cycle

of average productive waters remain in the range of 0.2 to 0.5 mg/l.

5.1.1 Plant Uptake

All plants can use nitrate and ammonium nitrogen, and some plants can use elemental nitrogen. Phytoplankton can absorb large amounts of ammonia and they are a dominant factor controlling the level of ammonium in pond water. In plant, nitrogen is reduced to ammonia and combine with organic carbon to form amino-acids. These amino acids are then combined to form protein. Plants may be consumed by animals or they may die and become detritus.

5.1.2 Animal Utilization

Animals eat plants, detritus, other animals or some combination of these substances. Animals need their nitrogen in the form of amino acids or protein. When animals take in food, part of the organic nitrogen in the food is assimilated and converted to animal protein. The remainder is expelled as organic nitrogen in faeces or as ammonia. When fish are harvested, the nitrogen that they contain is removed from the pond. Other animals are consumed by predators or they die and become detritus.

5.1.3 Degradation of Organic Matter

Some of the dead organic matter (detritus or smaller particles of material in soil and water) is consumed directly by animals. Ultimately, most dead organic matter becomes substrate (food) for microbial organisms (bacteria actinomycates, and fungi). Manures added to ponds also decomposes. Factors affecting the rate of organic matter decomposition are temperature, pH and the nature of the organic matter. Micro-organisms capable of degrading organic matter are present in all ponds. Their abundance will increase when organic matter is increased. Temperature in tropical fish ponds is ideal for growth of microorganisms. Degradation of organic matter proceeds most rapidly at pH 7 to 8. Therefore, in acid ponds, organic matter tends to accumulate unless lime is applied to improve pH. Organic matter that has a high concentration of nitrogen relative to its concentration of carbon (low C/N ratio) will decompose faster than a material with a high C/N ratio. Additionally, more nitrogen will be released to the environment as ammonia when a substance with a low C/N ratio decomposes.

Bacteria contain about 10 per cent N and 50 per cent C on a dry weight basis. They have a carbon assimilation efficiency of about 5 per cent. This means that for every 100 g of substrate carbon, bacteria will convert 5 gm of the carbon in the substrate to carbon in bacterial tissue. Bacteria are roughly 50 per cent carbon, so 100 g of substrate carbon will yield 10 g of bacteria. A 10 gm mass of bacteria would contain 1.0 g of nitrogen.

If an organic material is very low in nitrogen, there will not be enough nitrogen in it to effect complete decomposition. In this event, bacteria will remove nitrate and ammonia from pond water for use in decomposition of matter. Removal of nitrogen from the environment by bacteria degrading nitrogen-deficient organic matter is called immobilization of nitrogen.

5.1.4 Nitrification

The ammonia released to the pond water by decomposition may be used again by plants, or it may be nitrified to nitrate by chemoautotrophic bacteria. Oxidation of ammonium to nitrite by bacteria of the genus Nitrosomonas is the first step in nitrification.

$$NH_4^+ + O_2 \rightarrow NO_3^- + 2H^+ + H_2O$$

In the second step, nitrite is oxidised to nitrate by bacteria of genus nitrobacter.

$$NO_2^- + O_2 \rightarrow NO_3^-$$

These bacteria use the energy released in the oxidation of ammonium and nitrate to reduce carbon dioxide to organic carbon. In other words, these organisms can produce organic matter by a non-photosynthetic pathway. Of course, the amount of organic matter produced by nitrification in ponds is minute compound to the amount produced by photosynthesis. Nitrification is a significant source of acidity in ponds and is also important in reducing the degree of ammonia build up.

5.1.5 Denitrification

In absence of oxygen, many microorganisms can use nitrate or other oxidised nitrogen compounds as source of oxygen and as electron an hydrogen acceptors in respiration. This heterotrophic process is termed denitrification because gaseous forms of nitrogen are released as metabolites and lost from the pond. For example nitrate can be further reduced to nitrite and nitrite can then be further reduced to nitrous oxide and finally nitrous oxide can be reduced to gaseous nitrogen. Physiologically the process can best be defined as nitrate respiration. Denitrification occurs in pond soils where DO concentrations are low. It is the major loss of nitrogen from ponds.

5.1.6 Ammonia Voltalization

Some ammonia is lost from pond water directly to the air when the pressure of ammonia gas in water exceeds the pressure of ammonia in the air. This process is most important at pH 9 and above. The significance of ammonia voltalization in ponds is unknown but such high pH is never encountered in fish ponds.

5.2 Phosphorus

Phosphorus concentrations in unaltered pond water are seldom great enough to produce adequate phytoplankton for good fish production. Thus, phosphorus is very often introduced into ponds in the form of fertilizers and manures. The phosphorus contain in manure is released to the water only after decomposition of manure.

5.2.1 Fate of Phosphorus in Ponds

The phosphorus cycle in a pond is illustrated in Figure 5. A high concentration of phosphate will remain in the water for a few hours or days after the addition of phosphate as a chemical fertilizer in ponds and the concentration will quickly decline to the pre-treatment level. Some phosphorus loss from the water may be attributed to uptake by plants and bacteria. Heavy phytoplankton blooms can absorb large amounts of phosphorus. Much phosphorus will, however, be absorbed by the pond soil. Even that phosphorus initially absorbed by the phytoplankton will eventually be mineralized from organic matter and enter the mud.

The amount of phosphorus entering ponds from natural sources, including release from the soil, is usually rather small even in highly productive ponds. Phosphorus must be applied in fertilizers or manures to maintain productivity. Phosphorus in manures is released when manure is degraded by bacteria. The amount of phosphorus harvested in fish usually represents less than one third of the phosphorus added to ponds to stimulate fish production. Nevertheless, fish harvest is the biggest loss of phosphorus from a pond eco-system. Most of the added phosphorus remains in the pond as insoluble phosphate compounds in the soil. Unfortunately, the soil phosphorus is not highly available to plants in the pond.

In summary, phosphorus must continually be applied to ponds to promote fish production. Some of this phosphorus is absorbed by plants and passes through the food chain to fish ultimately, the largest part of the added phosphorus becomes soil phosphorus that is not highly available.

5.2.3 Reactions with Muds

Inorganic phosphorus in soils or muds occurs in several forms: Calcium, iron and aluminium phosphate. In acidic muds, aluminium, iron occurs at fairly high concentration and reacts with phosphate to form highly insoluble aluminium phosphate according to the general reaction.

$$Al^{3+} + H_2PO_4^- \rightarrow AlPO_4 + 2H^+$$

At the same pH, there are several orders of magnitude more aluminium ion than ferric iron in aerobic mud. Therefore, phosphate first reacts with aluminium, but the existence of iron phosphate in mud suggests that some aluminium phosphate is transformed to iron phosphate.

When muds become anaerobic, iron phosphate dissolve, and anaerobic water at the pond bottom may be high in phosphate. Nevertheless, when the water becomes aerobic again, the iron phosphate precipitates.

As pH of muds increases, concentration of Al^{3+} deceases so that less phosphate precipitates as aluminium phosphate. Somewhere between pH 5.5 and 6, the precipitation of Aluminium phosphate ceases to be an important factor. However, as muds increase in pH, the concentration of calcium increases and calcium phosphates precipitated when phosphorus is added. Over time, calcium phosphate is transformed to a highly soluble mineral apatite (rock phosphate). Where pH and calcium concentration are high, apatite may precipitate directly from the water.

Muds also contain organic phosphorus. Microbial decomposition of organic matter release phosphate which will participate in reactions with iron, aluminium and calcium.

As phytoplankton can quickly absorb phosphate from the water, so a large proportion of the phosphorus applied to a pond may enter phytoplankton, cells and promote growth. Nevertheless, phytoplankton cells are consumed or die, and the phosphorus from the phytoplankton eventually finds its way into the mud. In general, pond mud is a sink rather than a source of phosphorus.

Chapter 6
Toxic Metabolites

As a result of metabolic activity by organisms in ponds, carbondioxide, ammonia and hydrogen sulphide some times may reach harmful concentrations.

6.1 Carbondioxide

High concentrations of carbondioxide can be tolerated by aquacultural species, although fish are known to avoid CO_2 concentrations as low as 5 mg/l. Most aquacultural species will survive in waters containing upto 60 mg/l CO_2, provided DO concentrations are high. When DO concentrations are low, the presence of appreciable CO_2 hinders uptake of oxygen. Unfortunately, CO_2 concentrations normally are high when DO concentrations are low. This results because carbondioxide is released in respiration and utilized in photosynthesis. Dissolved oxygen concentration declines when photosynthesis is not proceeding as rapidly as respiration; thus CO_2 accumulates, because it is not removed for use in photosynthesis.

Because of the necessity of light for photosynthesis, CO_2 concentrations increase at night and decrease during day. High concentrations of CO_2 also occur in ponds during cloudy weather and following die-offs of phytoplankton or of filamentous algae.

It is seldom practical to remove CO_2 from pond waters. However, sometimes it is necessary to remove CO_2 from tanks or other containers in which fish are reared or held. Removal may be affected by applying calcium hydroxide $Ca(OH)_2$ or calcium oxide (CaO). To remove Img/l. CO_2, the application rates based on weight relationship among reacting compounds, are 0.64 mg/l for CaO and 0.84 mg/l for Ca $(OH)_2$. However, because of the low solubility of CaO and $Ca(OH)_2$ 1.5 to 2.0 times the calculated amount should be used.

6.2 Ammonia

Ammonia reaches pond water as a by-product of metabolism by animals and by decomposition of organic matter by bacteria. In water, ammonia nitrogen occurs in two forms, un-ionized ammonia (NH_3) and a ammonium ion (NH_4^+), in a pH and temperature dependent equilibrium.

$$NH_3 + H_2O = NH_4^+ + OH^-$$

As pH rises, un-ionized ammonia increases relative to ammonium ion. Water temperature also causes an increase in the proportion of un-ionized ammonia, but the effect of temperature is less than that of pH. Analytical procedures for ammonia nitrogen measure both un-ionized and ionized ammonia. Percentages of un-ionized ammonia at different temperature and pH values are available in the literature.

The toxicity of ammonia, nitrogen to fish is attributed primarily to the un-ionized form. As ammonia concentration in water increases, ammonia excretion by aquatic organisms diminishes, and levels of ammonia in blood and other tissue increase. The result is an elevation in blood pH and adverse effects on enzyme-catalyzed reactions and membrane stability. Ammonia increases oxygen consumption by tissues, damage gills, and reduces the ability of blood to transport oxygen. Disease susceptibility also increases in organisms exposed to sublethal concentrations of ammonia.

The tolerance of aquatic organisms to ammonia varies with species, physiological condition, and environmental factors. Lethal concentrations for short term exposure (24 to 72 hours) are between 0.4 to 2.0 mg/l of un-ionized ammonia. Percentages of un-ionized ammonia at 30°C and different pH values and concentrations of total ammonia nitrogen necessary to give 0.4 mg/l un-ionized ammonia under these conditions are tabulated in Table 5.

Ponds seldom contain more than 2 or 3 mg/l of total ammonia nitrogen. Obviously, ammonia toxicity will be a greater problem at high pH than at lower pH. It is difficult to evaluate ammonia concentrations in ponds. Because of the daily cycle in pH, un-ionized ammonia concentrations change continuously. Ammonia toxicity in fish usually expressed by reduced growth rate instead of mortality.

High ammonia concentrations are most common in ponds with high feeding rates. The only feasible means of reducing ammonia concentration is through exchange of water.

Table 5

pH	*Un-ionized Ammonia %*	*Concentration of Total Ammonia Nitrogen to give 0.4 mg/l NH_3*
7.0	0.81	49.38
7.5	2.56	15.62
8.0	7.52	5.32
8.5	20.73	1.93
9.0	44.94	0.89
9.5	71.43	0.56
10.0	88.80	0.45

Unionized ammonia concentration in the range of 0.02 to 0.05 mg/l is considered safe for many tropical freshwater fishes. Whereas in the range of 0.05-0.4 mg/l has sublethal effect depending on species and in the range of 0.4 to 2.5 mg/l become lethal to many fishes.

6.3 Hydrogen Sulphide

Under anaeobic conditions, certain heterotrophic bacteria can use sulphate and other oxidised sulphur compounds as terminal electron acceptors in metabolism and exerete sulphide as illustrated below:

$$SO_4^{2-} + 8H^+ \rightarrow S^{2-} + 4H_2O$$

Sulphide is an ionization product of hydrogen sulphide and participates in the following equilibria:

$$H_2S = HS^- + H^+$$

$$HS^- = S^{2-} + H^+$$

The pH regulates the distribution of total sulphide among its forms (H_2S, HS^-, and S^{2-}). Unionized hydrogen sulphide (H_2S) is toxic to aquatic organisms, the ionic forms have no appreciable toxicity. Analytical procedures measure total sulphide. Values given below show the percentages of unionized hydrogen sulphide at different pH values at 25°C which decreases as the pH increases.

Table 6

pH	*Hydrogen Sulphide (%)*
5.0	99.0
5.5	97.0
6.0	91.1
6.5	76.4
7.0	50.6
7.5	24.4
8.0	9.3
8.5	3.1
9.0	1.0

Concentrations of 0.01 to 0.05 mg/l of H_2S may be lethal to aquatic organisms. Any detectable concentration of hydrogen sulphide is considered undesirable. The presence of hydrogen sulphide may be recognized without water analysis, for the "rotten-egg" smell of hydrogen sulphide is detectable at very low concentration.

If water contains hydrogen sulphide, water exchange will reduce its concentration. Application of lime to raise the pH of the water will reduce the proportion of the total sulphide that is comprised of hydrogen sulphide. The concentration of hydrogen sulphide in the range of 0.01 to 0.5 ppm is lethal to fish and any detectable limit cause stress to fishes. Prawns lose their equilibrium at a concentration of 0.1 to 0.2 mg/l and create sub-lethal stress and at a concentration of 3 mg/l prawns die instantly.

Chapter 7
Soil

Soil is a dynamic natural body developed from natural substances as a result of natural process and is an integral component of pond. A section downward of the soil is called the profile (Figure 3) while each of the individual layer depending on depth exhibiting variation in texture, colour and structure is known as the horizons which reflect the soil formation process (Figure 3). Further soil is considered to be a three phase system consisting of solid, liquid and gas in addition to a living phase comprised of varied population of organisms (Figure 3).

The solid phase contains both mineral and organic matter and constitute only fifty per cent of the soil volume. The proportion of each of the mineral and organic matter varies greatly in different soils. The mineral matter of various soils range from 99 per cent in sandy soils and 95-97 per cent in many productive soils of intermediate texture. In

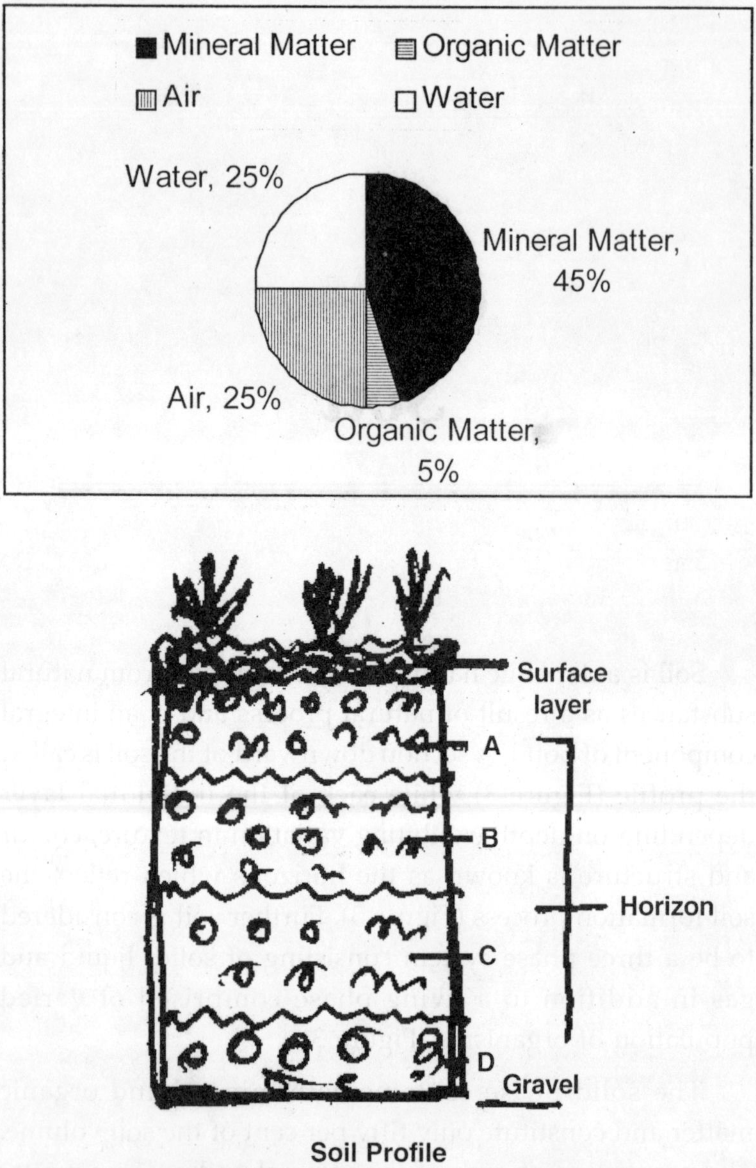

Figure 3: Percentage of Different Phases of Soil and Soil Profile

well drained coloured soils, the percentage drops to about 92 per cent and 80-60 per cent in poorly drained soils. Peat or muck soils (organic) may contain 60 per cent to as little as 5 per cent less of mineral matter. The organic matter content in well drained and upland mineral soil seldom contains more than 10 per cent organic matter and may contain even less than 1 per cent in poor sandy soils while the organic matter content of muck and peat soils remain from 20-50 per cent and even sometimes more than 50 per cent respectively.

The mineral matter of soil is one of the most important fraction and contains most of the nutrient element present in soil in inorganic form/combinations though some of the elements also remain in organic matter fraction of soil/as organic compounds/inorganic matter fraction of soil.

The mineral matter comprises of particles of different sizes varying from 0.2mm to less than 0.002 mm in diameter are known as S and, silt and clay according to the size of the particles. Particle size greater than 0.2mm to 2.0 mm in diameter is termed as gravel. All these factors to gather represent the texture of the soil and on the basis of the percentages of individual particle sizes influence the water holding capacity and also reflects the nutrient status of the soil.

The clay fraction is considered to be the most important among the different fractions of soil and contains most of the nutrient present in soil. The colloidal nature of clay particles exhibits all the colloidal properties such as exchange phenomenon, adsorption of nutrients besides water retention capacity of the soil which are of great relevance on availability of nutrients and suitability of soil for construction of fish ponds.

Besides, mineral matter, the organic matter content of soil are also colloidal in nature and exhibits all colloidal properties similar to clay particle and the nutrient remain mostly in organic combinations.

The nutrient element present in soil either, in the mineral or organic matter fraction remain in three different available forms as (*i*) readily available (*ii*) moderately available and (*iii*) slowly or difficully available forms. A constant transformation of nutrients elements from one form to another occurs in soil and the main direction of transformation depend mainly/largely on the presence of relative proportions of the various forms.

The liquid phase consists of soil moisture and water and varies in volume with the moisture content of the soil and 25 per cent of the volume consists of soil solution. Soil water is held in the soil pore space with varying degrees of tenancity depending on the amount of water present and together with the dissolved salts. Soil water makes up the so called soil solution which is important as a medium for supply of nutrients. In dry soil the concentration of soluble salts in solution usually, ranges from 100 to 1000 ppm and the soil becomes toxic when the concentration of dissolved salt remain 2000 ppm.

The amount of soil air varies with the amount of water present in soil and often occupies about fourth of the soil volume having an approximate composition of nitrogen (80 per cent) and the remaining 20 per cent mainly being oxygen including 0.01 per cent carbon dioxide in well drained soils. In waterlogged or submerged soils, toxic concentrations of methane, hydrogen sulphide and other substances develop under anaerobic conditions due to undesirable type of bacterial activity and also due to restriction on the escape of carbon dioxide and entry of fresh air in to soil.

7.1 Acid Sulphate Soil

Acid sulphate soil from mine spoils and coastal mangroves tide swamps contains high level of pyrites (FeS_2 1-6 per cent). As long as sediments containing pyrites are submerged and anaerobic they remain in reduced condition and change little. But on exposure to air, oxidation results with the formation of sulphuric acid.

Sulphuric acid reduce the pH of water when pond is filled with water. The problem with acid sulphate soil usually originates in pond dykes as pond bottom are usually flooded and remain anaerobic thereby prevent the formation of sulphuric acid which occur during dry period and enters pond with run off waters after rains. Acidity on dykes can be controlled by liming (0.5 to 1.0 kg/sqm) and with the establishment of acid resistant grass cover. But such practices are considered to be expensive and the results remain unpredictable.

7.2 Saline Soil

Saline soil contains a sufficient concentration of neutral soluble salts which adversely affect the growth of most plants. Less than 15 per cent of the Cation Exchange capacity of these soils are occupied by sodium ions and the pH usually remain below 8.5 as the soluble salts are mostly neutral.

These soils are also known as "White Alakali" soils due to surface incrustations if present is light in colour. The excess soluble salts are of mostly chlorides and sulphates of sodium, calcium and magnesium and can be leached out of soil without any appreciable change in soil pH which is of vital importance for management of such soils.

7.3 Saline-Alkali Soil

These soils contain appreciable amount of neutral soluble salts and enough adsorbed sodium ions. More than 15 per cent of the total exchange capacity of such soils are occupied by sodium and the pH usually remains below 8.5. But leaching of these soils markedly raise the soil pH due to the removal of the neutral soluble salts. After the removal of neutral soluble salts the exchangeable sodium present readily hydrolyses and thereby increase the OH ion concentration of the soil solution.

7.4 Non-Saline-Alkali Soil

Such soils do not contain any great amount of neutral soluble salts and the adverse effect on plants is mainly due to toxicity of Na^+ and OH^- ions. The sodium percentage usually remains more than 15 per cent of the total exchange capacity of these soils and is free to hydrolyse due to low percentage of neutral soluble salts and thereby often raise the pH to 10.0. The physical condition of such soils mostly remain in an unsatisfactory condition and readily change to alkaline soil. The extreme alkalinity resulting from the presence of sodium carbonates, the surface of alkaline soils usually become discoloured by the disperse humus and the term "black alkali" is used for such soils.

Chapter 8
Characteristics of Pond Soil

Soil is an integral component of pond environment and has several roles in Biological production of fish ponds, in supplying nutrients, mineralisation of organic bottom deposits and in the storage and release of nutrients to the water besides providing shelter and food for dwelling organisms.

Pond soil is formed under very specific conditions and differs from upland soils. It remains permanently waterlogged or submerged conditions though some analogy exists between pond soil and waterlogged rice soils. In ponds, the suspended and dissolved particles of both inorganic and organic materials entering into the pond environment from different profiles, through run off and planktonic organisms both living and dead produced in the system get sedimented and precipitated at the pond bottom over the parent soil. Thus under altered physico-chemical conditions these substances undergo various chemical and bio-chemical changes. As a result the characteristics of bottom soil is gradually modified with the advancement of time depending on climatic and environmental conditions to form the pond mud over the parent soil.

Under submerged conditions pond soil consists of two biologically significant layers. The loose aerated upper layer remaining in contact with water is a mixture of collidal mineral and organic matter formed by preciptation and sedimentation of both decomposed minute plant and animal organisms. The anaerobic lower layer contains mineral matter of varying composition depending on the parent rock from which the soil is derived. The proportion of these two layers regulate various physical, chemical and biological processes in soil like release of nutrients, adsorption of ions and bacterial decomposition of organic matter essential for biological production.

Nutrients required for biological production are released from soil through various biochemical processes to the water phase and vice-versa are regulated/influenced by these two layers and efficient transfer of nutrients the thickness of the upper layer should remain within a few millimeters.

But such a condition seldom exists/encountered in fish ponds even under normal conditions as the thickness of the bottom sediment tend to increase over time from negligible thickness in newly dug ponds to more than a meter in much older ponds due to increase in organic matter content of the bottom sediment.

Moreover, the increase/such increase in organic matter content of bottom sediment remains comparatively less/low in unmanaged ponds compared to well managed ponds. Further organic matter content of bottom sediment is one among the various factors which effect the productivity of fish ponds considerably. Accordingly, in newly dug ponds productivity remains low and gradually increases with the increase in organic content of the bottom sediment and ultimately become unproductive when the organic matter content of the bottom sediment exceeds more than 4 per cent.

8.1 Texture

The texture of a soil refers to the proportion of gravel, sand, silt and clay particles present in the soil. Particle size analysis of soil provides the percentages of sand, silt and clay from which a soil texture name *e.g.* sandy loam, clayey loam, loam etc. may be assigned. The percentages of different particle sizes present in soil is of great significance in indicating the water holding capacity of pond besides influencing the dosages of lime and indicate the suitability on the use of fertilizers or manures or a combination of both for management of fish ponds. Moreover, a knowledge on the percentage of clay present/content in a soil is essential as the clay is the reactive fraction. Soil also contains organic matter, and organic matter like clay is also highly reactive.

8.2 Cation Exchange

Colloidal particles of organic matter and clay minerals in pond soil have negative charges and attract swarms of cations. An equilibrium exists between concentrations of cations in water surrounding soil particles and amounts of cations adsorbed on soil particles (Figure 4).

Cations on soil colloids and in surrounding water are known as exchangeable cations and the sites of adsorption on colloids are known as exchange sites. Some cations are held more tightly than others to colloids.

The quantity of cations which can be adsorbed on soils is called the cation exchange capacity (C.E.C). The C.E.C. is measured in milliequivalents of cations per 100 gm. dry soil. The larger the C.E.C. of the soil, the greater is the ability of the soil to exchange and hold ions. The C.E.C. of pond mud ranges from less than 1.0 meq/100 gm. to more than 100 meq/100 g. The C.E.C. increases as the percentages of clay, organic matter or both increase. Some types of clays have greater exchange capacities than others.

8.3 Acidity

The cations adsorbed on exchange sites in soils are acidic ($A1^{2+}$, Fe^{3+} and H^+) or basic (Ca^{2+}, Mg^{2+}, K^+, Na^+ and NH_4^+) the fraction of the total exchange capacity occupied by acidic ions is called the base unsaturation. In most soils, there will be very little H^+ or Fe^{3+} ion on exchange sites. The primary acidic ion is $A1^{3+}$. The acidic reaction of Aluminium may be visualised as follows:

$$A1\text{-}Soil = A1^{3+} + 3H_2O = A1(OH)_3 + 3H^+$$

As the base unsaturation of a soil increases, the amount of $A1^{3+}$ available to react with water and form hydrogen ion increases. Therefore soil pH decreases with increasing base unsaturation.

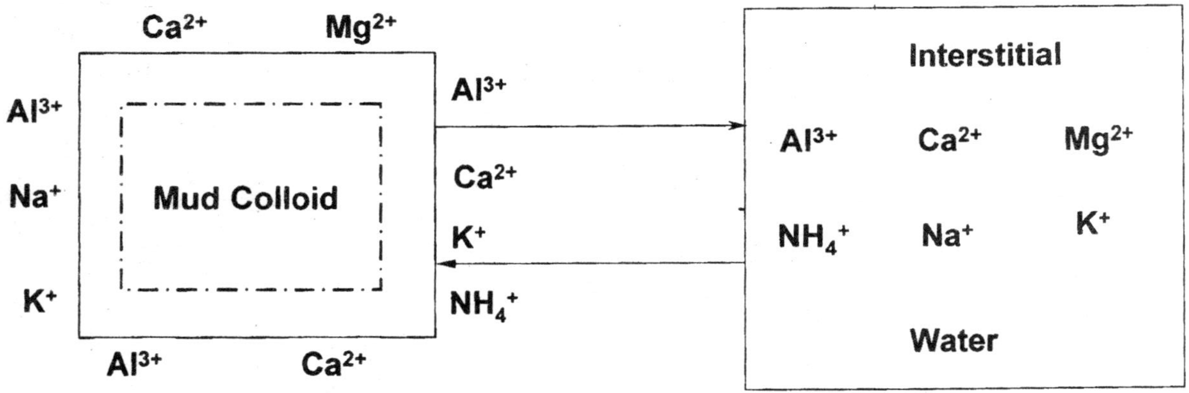

Figure 4: Nutrient Exchange Between Soil Colloid and Interstitial Water

The calcium carbonate react with hydrogen ions and neutralizes them. This lowers the concentration of hydrogen ions in solution and more aluminium ions are released from the soil. Aluminium ions released from the soil are replaced by calcium ions resulting from the neutrilization of hydrogen ions by calcium carbonate. The end results are: aluminium is removed from the soil and precipitated as aluminium hydroxide; calcium replaces the aluminium on the soil; the base unsaturation of the soil decreases; the pH of the soil increases.

8.4 Oxidation-Reduction

Under restricted movement of water through muds, microbial activity quickly reduces the oxygen concentration in the water within the mud (interstitial water) and in the water lying just above the mud surface. As the oxygen concentration drops, the oxidation potential falls and various compounds are reduced. Although some of these reduction may occur spontaneous in the absence of oxygen, most are carried out and accelerated by microorganisms. Some typical reductions that occur in pond muds are:

NO_3^- to NO_2^-	Fe^{3+} to Fe^{2+}	SO_4^{2-} to H_2S
NO_2^- to NH_3	Mn^{4+} to Mn^{2+}	

The degradation of organic matter in the mud cause the low DO condition and the continued degradation of the organic matter results in the reduction of the inorganic substances. Thus, organic matter is the source of reducing power that often leads to high concentrations of nitrite, ammonia, ferrous iron (Fe^{2+}), diavalent manganese ion and hydrogen sulphide in pond muds.

In absence of oxygen, micro-organisms do not degrade organic matter to carbondioxide and therefore, organic matter accumulates at a greater rate where muds are anaerobic.

The iron reaction in water provides a means of determining whether the surface layer of a mud is anaerobic. In the absence of oxygen ferric iron (Fe^{3+}) is converted to Ferrous iron (Fe^{2+}). Ferrous iron is black in colour. Therefore, when the surface of the mud is black, anaerobic condition exists and the surface remains brown or the natural soil colour, when oxygen is present. A break through the surface of an aerobic mud will expose the anaerobic conditions and black in colour of the deeper layers. It is highly desirable to maintain dissolved oxygen in the upper layer of mud. Fish food organism that live in mud require oxygen and the presence of oxygen in the mud prevents formation of noxious reduced substances.

8.5 Muds and Fish Production

Although relatively little is known about the relationships between chemical and physical properties of muds and fish production, there has been one excellent study that suggests muds are extremely important to fish production. The results are summarized in Table 7.

Inspite of the correlations between soil properties and fish production, good production of fish cannot be sustained without application of manures, fertilizers, feeds or some combination of these, since soil and water are mostly remain deficient in major nutrients essential for production.

Table 7

Variable	Range	Fish Production
pH	5.5	Low
	5.5–6.5	Average
	6.5–7.5	Optimum
	7.5–8.5	Average
Available Phosphorus	3 ppm	Low
	3–6 ppm	Average
	6 ppm	Optimum
Available Nitrogen	25 ppm	Low
	25–50 ppm	Average
	75 ppm	Optimum
Organic Carbon	0.5 per cent	Low
	0.5–1.5 per cent	Average
	1.5–2.5 per cent	Optimum
	> 2.5	Declining

Chapter 9

Significance of Fish Pond Management

The mineral fraction of soil contains various elements and exhibits a wide variation in their amount from one region to another depending on soil characteristics and from surface to subsurface soil. Soil is the major source of various elements recognised as essential for healthy development of plants, growth of animals and micro-organisms, except carbon, hydrogen and oxygen which are obtained from the atmosphere. Moreover, the requirement of some of the elements are comparatively greater than other elements compared to other elements. Accordingly these elements are known as major or macro elements and minor or micro or trace elements according to their requirements. Nitrogen, phosphorus and potassium are considered as primary nutrients. According to their significance in production processes nitrogen, phosphorus and potassium are considered as primary nutrients while, calcium and magnesium are secondary nutrients and Fe, Cu, Mn, B, Co, Zn, Mo, S and Chloride fall in to the category of minor or trace elements. In addition to these elements the importance of selenium and vanadium have been recognised for plants and animals in recent years. Among these elements nitrogen, phosphorus and potassium are more relevance in direct involvement in regulating important biological production process compared to other elements while calcium and magnesium influence in maintaining/creating healthy environmental conditions besides better availability of nutrients for biological production. Whereas, exact mechanism of utilization of various micro-nutrients in biological production are not well known though several trace elements have been found to stimulate the growth of fish and also effective for certain enzyme systems.

Besides, significance of various elements in biological production loss of water through seepage and maintenance of a favourable pond environment for better availability of nutrients, reduction of toxic or harmful effects of some elements including bacterial activity are also great relevance in biological production and thereby necessitate adoption of suitable management practices to achieve the desired objective.

9.1 Seepage

Loss of water is a natural phenomenon in fish ponds and occurs through both upward movement of water known as evaporation while the downward movement through the soil profile is termed as seepage. Variation in the water temperature cause loss of water through evaporation whereas seepage loss is attributed to larger pore space between the different soil particles and lack of compactness between the soil particles. Though pond exhibits more or less loss of water through seepage to certain extent but excessive loss of water is a matter of concern in fish ponds. In normal ponds loss on seepage remain even less than 1.0 cm/day depending on the texture of pond soil, whereas seepage in most ponds remain in the range of less than 0.1 cm to 1.0 cm/day depending on the porespace between the different particle size of sand, silt and clay present in the soil which ultimately influence the soil compactness and thereby cause a wide variation in the rate of seepage in soils.

The size of the particles present in soil vary from less than 0.002 mm in diameter to more than 0.02 mm in diameter and their percentage present in the soil influence the rate of seepage in soil. According to the size, particles are divided into sand (0.20-0.02 mm), silt (0.02-0.002 mm) and clay (less than 0.002 mm) in diameter while in the range of 0.20-2.0 mm in diameter are known as gravel. The percentage of all these fractions together present in soil is known as the texture of the soil. Thus, sandy soils are generally highly porus, while clay soils are more compact and impervious in nature and is undesirable/unsuitable for fish ponds.

Due to excessive loss of water and availability of nutrients respectively. Loam soils with a balanced percentage of sand (23-52 per cent), silt (28-50 per cent) and clay (7-27 per cent) is most suitable for fish pond. However, total prevention of seepage is never feasible but could be reduced considerably or to desired level through suitable management measures.

9.2 Soil Reaction

Soil reaction usually denoted as pH is of much significance in affecting availability of nutrients, toxic effect of various elements bacterial activity, degradation of soil organic matter and buffering in soil. So management of soil reaction is essential to maintain a favourable soil reaction to counteract the adverse effect of soil reaction on both sides of neutrality for maintaining a favourable pond environment.

9.3 Nitrogen

Nitrogen being a constituent of protein and chlorophyll is essential for the formation of living matter, synthesis of nitrogenous compound, increasing the percentage of protein, vegetative growth of aquatic plants and governs a considerable degree on utilization of phosphorus, potassium and other constituents, acts as a resistance to diseases and synthesized into cell content of soil micro-organisms. Thus, the amount of nitrogen withdrawn is fairly large compared to its presence in the system.

Soil is the main source of this element and also present as a constituent of soil organic matter in fish pond besides enrichment of this element to certain extent from atmosphere through electric discharge and biological fixation. Degradation of organic matter also release various nitrogenous compounds viz, ammonium, nitrate and nitrite depending on pond environmental conditions including loss of gaseous nitrogen under anaerobic conditions and through downward movement of soluble nitrates. The relationship between the different forms of nitrogen is regulated through the operation of nitrogen cycle energised through bacterial activity (Figure 2) and a dynamic equilibrium is always maintained between these different forms in pond environment.

The presence of this element in both soil and water usually remain more than both phosphorus and potassium. The total nitrogen content of most surface soils remain in the range from 0.02 to 0.4 per cent and decrease with depth. A major part of the total N content of 92-96 per cent remain in organic combinations. The soil organic matter contains about 5 per cent nitrogen. The nitrogen content of soil organic matter is about 5 per cent and approximately 98 per cent of N occurs in organic form, while the inorganic form rarely exceeds 2 to 3 per cent of the total N content of soil.

The available N content is considered to be more important than the total nitrogen content of soil in deciding the deficiency of this element in pond soil for biological production. The available N content of fish pond soils normally vary from less than 250 ppm to more than 500 ppm in newly dug and old ponds respectively,while the dissolved inorganic nitrogen content of water remain within 0.04 to 0.2 ppm in most fish culture ponds and may exceed more than 0.2 ppm under exceptional conditions.

9.4 Phosphorus

Phosphorus ranks next to nitrogen in importance is considered to be a key element in biological production and very often acts as a limiting element in aquatic production. It is concerned in structural compounds nucleic acids, reproductive processes including conservation and transfer of energy in metabolic reactions including growth and synthesis of protein. Phosphorus is also essential for assimilation of nitrogen into cellular matter respiration and is responsible for cell division, fat and albumin formation, counteract the ill effects of excess nitrogen including resistance to certain diseases.

The phosphorus content of most soils is rather low and 30-40 per cent of the total phosphorus remain in organic combinations in microbial tissues in the form of nucleic acids and phospho-proteins, while 75 to 80 per cent of the total phosphorus remain in organic matter content of soil. The total phosphorus content of some fish pond soil has been observed to be in the range from 0.02 to 0.088 per cent. Lack of this element and conversion or fixation on external application makes its importance doubly serious either due to conversion or fixation to unavailable form on external applications and may prevent other nutrients being acquired by plants. Further the reactive nature of phosphate ions render this element unavailable due to fixation or conversion of soluble phosphates into insoluble iron, aluminium and calcium phosphates under both acidic and alkaline conditions respectively. Thus, the form of phosphorus and conditions affecting their availability is important rather than the total phosphorus content of soil. Mineralization of organic phosphorus content of soil to a great extent is regulated by the physico-chemical properties of the soil and in most cases parallels to that of carbon and nitrogen while all the three elements are involved in the transformation of organic fraction of soil. Mineralization of organic phosphorus is more sensitive to changes in soil pH and an increase in soil reaction results in the rate of mineralization of phosphorus and even exceeds those of carbon and nitrogen.

Besides, physico-chemical characteristics, soil temperature, cultivation practices and nature of added organic residues also affect the mineralization of phosphorus. The extent of mineralization or immobilization largely depends on C:P ratio of the added residues. A wider C:P rate will immobilize a greater part of the phosphorus released while liberation of excess phosphorus will occur with a narrow ratio of carbon to phosphorus.

Unlike nitrogen, loss of phosphorus never occur in soil. The soluble inorganic phosphorus content of most fresh water fish pond remain as traces and seldom exceeds more than 0.1 ppm, whereas, the available phosphorus remain less than 20 ppm in most ponds and seldom exceeds greater than 60 ppm. In newly dug and old ponds, the available phosphorus contents in soil even remain less than 10 ppm.

9.5 Potassium

Potassium is another major cation ranks along with nitrogen and phosphorus of importance in aquatic production. It is essential in nutrition of plants, helps in the formation of protein, chlorophyll, acts as resistance to certain diseases, counter the ill effects of nitrogen, exerts a balancing effect on both nitrogen and phosphorus and stimulate the growth of aquatic plants. But the importance of this element in biological production is less obscured compared to nitrogen and phosphorus due to its easy availability and remaining higher than both nitrogen and phosphorus in most fresh water ponds. Moreover, micro-organisms also affect the availability of potassium through liberation of organically bound potassium. The total potassium content of some fish pond soil has been found to be in the range from 0.51 to 1.13 per cent.

Potassium occurs in soil in greater combination and only a fraction of 1 to 2 per cent remain in readily available form. The available form exhibits an equilibrium with relatively unavailable and slowly available form comprising of 90 to 98 per cent and 1 to 10 per cent of the total potassium content, respectively, which has been represented as follows :

Relatively unavailable K
(90–98 per cent of total K)

↙ ↘

Non-exchangeable K $\rightleftharpoons$ Exchangeable K-soil solution K
(Slowly available) (Readily available)
(1–10 per cent of total K) (1–2 per cent of total K)

In soil, the exchangeable form of potassium makes up readily all the available category varying from 155 to 370 ppm and only an insignificant amount exists in water soluble salts comprising of mostly, nitrates, sulphates and chlorides. Moreover, several soil related factors like texture, organic matter, pH, wetting and drying influence the availability of this element in soil. The readily available potassium of 1 to 2 per cent of the total K remain in solution while 90 per cent remain in exchangeable form.

The deficiency of this element never occur in pond except peat or muck soil. The dissolved potassium content remain more than 1.0 ppm in most freshwater fish pond.

9.6 Carbon

Carbon is another element of great significance in biological production. It is a primary constituent of carbohydrate. In ponds, chlorophyll utilizing the atmospheric carbon dioxide in presence of sunlight produces carbohydrates through photosynthesis which form the primary link in aquatic food chain and acts as a source of energy for various bacteria responsible for degradation of soil organic matter and transformation of nutrients. Thus, photosynthic activity converts inorganic carbon to organic carbon while the process is reversed through bacterial degradation of organic matter with the liberation of inorganic carbon as carbondioxide.

Both inorganic and organic carbon occur in pond as carbondioxide, bicarbonate, carbonate and organic matter respectively. A dynamic equilibrium exists between the different forms of inorganic carbon and thereby a wide fluctuation in total alkalinity of water is prevented in fish pond considered to be detrimental/harmful for aquatic animals.

$$CO_2 + H_2O \leftrightarrows H_2CO_3 \leftrightarrows CO_3^{2-} + H_2$$

Carbon present as organic matter in surface soil declines in sub-soil. Fish ponds very often remain unproductive due to lack of this element in suitable amount initially after construction of pond compared due to high organic matter content in older ponds. Thus productivity of pond remain low at both the extremes of this element in soil. Pond soil usually exhibit a wide variation in their organic carbon content depending on their age, culture practices etc. and remain less than 0.05 per cent in newly dug ponds to more than 2.5 per cent in much older pond. Under normal conditions, submergence of soil also enhance the organic matter contents of pond soil and the increase remain comparatively higher under various pond management practices, while dewatering of ponds cause a reduction in organic carbon content of soil.

9.7 Calcium and Magnesium

Calcium and magnesium are considered as secondary elements in pond management due to their importance in maintaining a favourable environmental conditions for growth of aquatic animal rather than biological production. Lack of these element make the soil acidic and thereby affect the availability of nutrients specially phosphorus required for biological production, enhance the toxic effects of some elements, hamper the activity of micro-organisms including buffering action in soil. The presence of calcium in soil is comparatively greater than magnesium in most soils and found to be in the range of 880 to 3870 ppm and 210 to 460 ppm in some fish pond soil, respectively.

9.8 Micronutrients/Trace Elements

Several trace elements *viz.* Fe, Cu, Mn, B, Co Zn, Mo, Si, I, F, Se and Va have been recognised essential for healthy development of animals, plants, micro-organisms and are considered to be effective in certain enzyme systems. However, the significance of some individual elements in biological production have been summarized as follows:

Iron and Copper have been considered in oxidation reduction processes in plants while copper is required for respiration and utilization of iron, nitrogen and its transformation in micro-organisms, oxidation reduction in plants, important reactions in plant metabolism while decrease in water adsorption, translocation of sugar, synthesis of protein contained in chloroplasts are effectively carried out in presence of boron. Whereas, Boron and cobalt are responsible in fixation of nitrogen and besides cobalt in production of vitamin B_{12} and growth of animals.

It is obvious from the term that the status of these elements in soil are much lower depending on the variation in soil characteristics compared to other elements present in soil. Among the various trace elements, some of them have been found to be much useful in short-term culture operations and plankton production.

Chapter 10
Management of Fish Pond

Soil being a natural body formed from natural substances through natural processes exhibits a wide variation in their physico-chemical characteristics from one region to another which cannot be altered to great extent. Thus, soils are very often found to be deficient in one or more nutrient element to meet the requirement of specific culture systems. Besides nutritional deficiencies, loss of water through seepage, soil acidity are the other factors, which affect the suitability of ponds for culture operations.

Pond management is related to regulate some environmental conditions so that they are within a desirable range for survival and growth of fish. In fish culture ponds, production is enhanced with the application of chemical fertilizers, organic manures and supplementary feed either alone or in combination. Where fertilizers are used, management usually involves manipulation of nutrient budgets with external application of nutrients either through chemical fertilizers or organic manures to increase plankton production for enhanced growth of fish. Some waters are too acidic and must be limed before fertilizers are used. Pond waters may be so turbid so that penetration of light in to the water medium is inadequate for fertilizers or manures to increase photosynthesis. Water exchange rates may be so high that fertilizers nutrients are flushed from ponds before they can exert a favorable influence on plankton production. Thus, the proper form, grade and amount of fertilizers or manures should be decided depending on the texture, nutrient status and reaction of the soil. Excessive application of nutrients through fertilizers or manures will encourage over abundance of plankton and may cause problem of dissolved oxygen whereas, insufficient application of fertilizers may result low levels of turbidity resulting in aquatic weed infestation.

A knowledge on deficiencies of nutrients and their magnitude including the texture of the soil and soil reaction is essential prior to their management. The informations are obtained from the examination of the relevant physico-chemical characteristics of bottom soil. Since the quality of water is considered to be more or less a reflection of the bottom soil due to the existence of a dynamic equilibrium between the two phases of pond environment so some informations of the nutritional deficiencies and soil reaction could be obtained through the chemical examination of water quality of fish pond.

Besides, manipulation of nutritional deficiencies of fish pond suitable management measures also essential to reduce the excessive loss of water due to seepage and also correction of the acidity of the soil prior to adoption of nutrient management in ponds.

10.1 Seepage

Excessive loss of water through seepage is considerably reduced with the application of higher dose of organic manures up to the extent of 20,000 to 25,000 kg cow dung/ha/yr in combination with lime at the rate of 200 to 300 kg as $CaCO_3$/ha/yr consecutively for several years to reduce the pore space between the soil particles. In certain cases, bentonite a synthetic clay is also used to reduce the seepage but such practice is considered to be rather expensive.

10.2 Liming

Liming in the first step of management of fish pond. Ponds are very often linked to raise the soil pH to a desired value *i.e.* neutrality to counteract the harmful affects of soil acidity. Liming is beneficial mostly in acid ponds with total alkalinity and total hardness remain below 20 mg/l. Liming materials are not fertilizers, but they will improve the response to fertilization or manuring in acidic ponds. The effects of liming on water quality are mentioned below.

1. Increase in soil pH.
2. Increase in alkalinity and hardness of water.
3. Flocculate suspended soil particle.
4. Increase the buffering capacity of the water.
5. Increase availability of carbon for photosynthesis.
6. Enhance bacterial activity in soil.
7. Increase availability of phosphorus.

Acting in combination with fertilization and manuring these improvements in water quality can lead to higher fish production.

The requirement of lime is made by a fairly simple laboratory procedure, for determining the lime requirement of pond soils. This procedure will work on all soils except acid-sulphate soils. When it is not possible in determining lime requirement, liming is done according to the Table 8.

Table 8

Soil PH		*Amount of $CaCO_3$ kg/ha*		
		Sandy Soil	*Medium*	*Clay Soil*
4.0-5.0	Highly Acidic	1000	2000	3000
5.0-6.0	Moderately Acidic	600	1200	1800
6.0-6.5	Slightly Acidic	500	1000	1500
6.5-7.5	Near neutral	200	400	600

Some authors recommend application of smaller amounts of liming materials (100 to 200 kg $CaCo_3$/ha)

Agricultural lime stone (pulverized calcium carbonate or dolomite), hydrated lime (calcium Hydroxide), and quick lime (calcium oxide) are the three most common liming materials used in ponds. Agricultural limestone is not harmful and it will not cause high PH in water and thereby extensively used in ponds due to several advantages over other liming materials. It can be applied to ponds even during the culture operations but the action of calcium carbonate is rather slow compared to oxides and hydroxides. In exceptional cases, calcium oxide may be used but application of calcium oxide on the dry pond bottom after draining the pond produce the best results.

The total amount of lime is applied either in one installment prior to the culture operations or in installments at an interval of three months. Application of lime is made either spreading it uniformly over the entire bottom of an empty pond or over the entire surface of a pond. Simultaneous application of both lime and chemical fertilizer should be avoided and application of fertilizers should be made preferably after a lapse of fortnight.

10.3 Fertilization

Based on the nutrient status of pond soil, the nutrients found to be deficient are corrected through external application of either chemical fertilizers or organic manures containing these nutrients.

Whereas, lack of information on the micro-nutrient status of fish pond soils, their deficiencies of individual elements for fish including interactions between these elements are the major constraints which has led to non-adoption of a suitable micronutrients management practices compared to other major nutrients in aquatic systems. Inspite of such constraints, application of manganese has been found to enhance/increased production of plankton in fish ponds while addition of Boron, Cobalt, Zinc and Molybdenum either alone or along with supplementary feed enhanced the survival and growth of carp spawn.

10.4 Chemical Fertilizers

Inorganic fertilizers are substances which contain nitrogen, phosphorus and potassium either alone or in combination are very often applied to fish ponds. The common fertilizers compounds used for fertilizing fish ponds are presented in the Table 9. When two or more of these materials are blended, the resulting mixture is called a mixed fertilizer while fertilizers containing all the three elements are known as complete fertilizers. Fertilizers are classified according to their nutrient content as nitrogen is expressed as N, Phosphorus is expressed as P_2O_5 and potassium as K_2O. In nitrogeneous fertilizer N is present as nitrate (NO_3), ammonium (NH_4) or urea $(NH_2)_2Co$. Phosphorus is present as $(H_2PO^2{}_4)$, while Potassium occurs as potassium in (K^+). Fertilizers containing individual nutrient element are rather suitable and used extensively for fertilizing fish ponds rather than compared to mixed fertilizers containing more than single or individual nutrient element.

Table 9: Approximate Grades of Common Fertilizers

Fertilizers	*Percentages*		
	N	*P_2O_5*	*K_2O*
Urea	45	0	0
Calcium Nitrate	15	0	0
Sodium Nitrate	16	0	0
Ammonium Nitrate	33	0	0
Ammonium Sulphate	21	0	0
Supper Phosphate	0	16	0
Triple Superphosphate	0	46	0
Mono ammonium phosphate	11	48	0
Diammonium Phosphate	18	48	0
Ammonium polyphosphate	10-13	34-37	0
Muriate of potash	0	0	60
Potassium sulphate	0	0	22

10.5 Organic Manures

Manures are animal excrements and other agricultural

wastes of largely organic matter but they contain some nitrogen, phosphorus and potassium. Dung of various animals *viz.* cow, buffalo, goat, poultry, pig and horse are very often used as organic manures in fish ponds. Among these manures, cowdung and the poultry manures are extensively used for manuring fish ponds.

Table 10: Composition of Fresh Manures of Selected Farm Animals and Oil Cakes

(Average composition)

Manure	*Moisture%*	*N%*	*P_2O_5%*	*K_2O%*
Cowdung	85	0.3	0.1-0.2	0.1-0.3
Poultry	72	1.0-1.8	1.4-1.8	0.6-0.9
Swine/Pig	82	0.5	0.5	0.45
Sheep & Goats	77	0.75	0.5	0.45
Farm yard Manure		0.50	0.2	0.50
Ground nut Cake		7.3	1.5	1.30
Mahua oil cake		2.5	0.8	1.80

Extensive research/studies has been conducted on the use of fertilizers and manures on fertilizing fish ponds as a practice of management.

Application of nitrogen in combination of phosphorus has been found to be more effective in enhancing biological production compared to either nitrogen and phosphorus alone. Moreover addition of a small amount of potassium to the combination of nitrogen and phosphorus further enhanced the production higher than the combination of nitrogen and phosphorus.

An effective schedule for fertilizing fish ponds with chemical fertilizer has been presented in Table 11.

Table 11: Fertilization Schedule for Fish Ponds

Nitrogen (N)				Phosphorus (P_2O_5)
Rate of Application of N (Kg/ha/Yr)	*Available N (Mg/100 gm)*	*Soil Type*	*Available P_2O_5 (mg/100gm)*	*Rate of Application of P_2O_5 (kg/ha/yr)*
150-200	50	High	6.0	75-100
200-250	25-50	Med	3.6	100-125
250-300	25	Low	3	125-150

Besides nitrogen and phosphrous a smaller amount of potassium at the rate of 40-60 kg. K_2O/ha/Yr is also applied along with nitrogen and phosphorus to achieve better yield either with the application N and or P alone or a combination of both. Moreover, organic manure at the rate of 10,000/Kg/ha/Yr. is also added when both inorganic and organics are used for management of fish ponds.

For efficient utilization of nutrients the total amount of nutrients are either added in the form of fertilizers or manures singly or incombination are applied in the ratio of 4:3:2:1 per quarter beginning from post monsoon to summer so as to cover the entire culture period/growing period. The amount for each quarter is further divided into split doses of equal amount and applied at fortnightly intervals for fertilizers and at monthly intervals in case of organic manures and alternately when both are used. Moreover application of fertilizers immediately after liming and also simultaneous application of both lime and fertilizers should always be avoided. In such cases fertilization should be undertaken after a fortnight of liming while both liming and manuring could be undertaken without any adverse effect.

The method of application of fertilizers and manures also influence the efficient utilization of nutrients added to fish ponds. For better utilization, fertilizers are applied in solution and broadcasted over the pond surface for immediate utilization by phytoplankton whereas organic manure are usually placed in heaps under water at different corners of the pond and the nutrients are released only after the decay of the added manures. Thus, the nutrients are released slowly and remain available for a longer duration compared to chemical fertilizers. In exceptional cases a part of the organic manures may be applied over the water surface after mixing with water to serve as a food for some aquatic animals.

Granular fertilizers when broadcasted over pond surfaces, settle to the bottom before they dissolve in water. In such cases phosphate is absorbed strongly by mud and much of it never reaches the water. In some countries liquid fertilizers have been found to be much more efficient than granular fertilizers are used because they will never settle to the bottom. The three common liquid fertilizer used in aquaculture are liquid urea (45-0-0), phosphoric acid (0-54-0) and ammonium polyphosphates (10-34-0 to 13-37-0). Liquid fertilizers are diluted with water prior to its application to pond surface otherwise will settle to the bottom due to heaviness.

As an alternative method of application, granular fertilizers are placed on an underwater platform so that water currents move the nutrients throughout the water after they dissolve. Sometimes fertilizers are put in porus bags and suspended in ponds so that nutrients are released into water as fertilizer dissolve. Either procedures prevents fertilizer/granules from dissolving while in contact with pond bottom soil. Nevertheless, regardless of the technique used for application of fertilizers, the phosphorus is eventually assimilated by plants, adsorbed by pond muds, or lost in outflowing water. Nitrogen is not strongly adsorbed by pond muds, but it is lost through denitrification and in outflow.

The recommended amount of nutrients required for application to ponds according to the table is reduced correspondingly taking into the account the addition of nutrients (N and P) from other sources, such as organic manures, when both inorganic are organics are used. Similarly, the amount of nitrogen added through supplementary feed in each quarter is also reduced from the amount of N required for application in that quarter and application of the balance amount of N as fertilizers are made when supplementary feed is provided and use of organic manure is avoided in such cases.

Water quality and the soil reaction affect the performance of fertilizers. In hard waters with high pH it is better to use ammonium phosphates than super phosphates or triple super phosphate. The performance of added nitrogenous fertilizers to fish ponds are also influenced by the reaction of soil. Accordingly, calcium ammonium nitrate is best for moderately acidic pond soil (pH 5.5 to 6.5), urea in neutral soil (pH 6.5 to 7.5) and ammonium sulphate for alkaline soils (pH-7.5 to 8.5). Among the phosphate fertilizers, single super phosphate is more preferred than triple super phosphate as a source of phosphorus for fertilizing fish ponds while muriate of potash is considered to be rather suitable compared to other forms of potassium as sulphates and nitrate for management of fish ponds.

Further texture of pond soil also influence the performance of added fertilizer and manures. Thus, chemical fertilizers are more suitable for clay soils and organic manures for sandy soils while a combination of both fertilizers and manures are suitable for loam soils.

10.6 Advantages and Disadvantages of Fertilizers and Manures

Chemical fertilizers containing single nutrient element are preferred ever other mixed or complete fertilizers containing more than single element for fertilizing fish ponds as the manipulation of nutrient composition is easier for application to ponds. Moreover, the higher and the fixed percentages of nutrients present in fertilizers a smaller amount of fertilizers are required for fertilizing ponds. Further, the higher/solubility of chemical fertilizers, the nutrients are released immediately after application to ponds. However, continuous use of chemical fertilizers specially nitrogenous fertilizers and potassium sulphate very often cause problems of soil acidity whereas application of either excessive amount of chemical fertilizers create problems of oxygen depletion due to over abundance of phytoplankton, while a combination of both chemical fertilizer and organic manure are sometimes used to stimulate zooplankton blooms in fish ponds.

Organic manures are composite in nature and the presence of most of the nutrient elements required for biological production make the organic manures more beneficial for manuring fish ponds. Moreover manures serve as a direct food for zooplankton and also provide a substrate for microbial growth. Thus, manuring is considered to be more effective for increasing zooplankton abundance in fish pond. Manuring also improves the soil structure and also reduce the loss of water through seepage to some extent.

The variable and low percentage of nutrients present in organic manures are the major disadvantages and thereby the requirement of manures for an equivalent amount of nutrients are comparatively higher than fertilizers. Further, the nutrients present in manures are not readily available after application to ponds unless manures are decomposed. The higher C/N ratio with low phosphorus content of organic manures make the decomposition of manures slow and thereby cause a progressive increase in organic matter content of bottom sediment. Fish production under such condition tends to decline over time in manured ponds due to accumulation of organic matter. Moreover, application of large amount of manures to ponds may cause depletion of oxygen to meet the demand of oxygen for decomposition of manure.

10.7 Aeration

The occurrence of oxygen-deficiency in fish ponds under normal pond environmental conditions and culture practices is never encountered and becomes a problem when the consumption of oxygen exceed than its availability in pond. In fish ponds, the dissolved oxygen content of water exhibits a diurnal variation due to variation in production and consumption processes occurring in the pond environment. As a result the dissolved oxygen content of water remains minimum at dawn and increases to attain maximum at noon and declines thereafter due to variation in temperature and photosynthesis activity (Figure 5). Moreover, the dissolved oxygen also shows variation between the surface and bottom layers of pond depending on the depth of water and remain less under normal conditions.(Figure 6).

In intensive culture systems with high density of stocking and provided with feed, fertilizers and manures to achieve higher yield and also in absence of water exchange very often cause depletion of oxygen in pond as consumption exceeds the amount of dissolved oxygen present in water. Under such conditions the uptake of feed is considerably reduced and thereby affect the growth of fishes and in extreme cases may cause mortality of fishes.

Under such conditions to overcome the problem of deficiency of oxygen, pond water is aerated with the operation of aerators to alleviate the dissolved oxygen content of water (Figure 7) and is essential when the concentration of dissolved oxygen fall between 2-3 mg/l and continued until there is no longer danger of depletion of dissolved oxygen. Moreover, aeration prevents stress and mortality in fish with occasional oxygen depletion. Such type of aeration is commonly known as emergency aeration and is seldom necessary except only between midnight and dawn during prolonged periods of cloudy weather or after plankton die-offs.

Aeration may be applied on a continuous basis to permit high production. Efficient use of aeration is to stock ponds at a modest rate and aerate every night from midnight to dawn to assure that the DO concentration always remain optimum. The duration of aeration and the number of

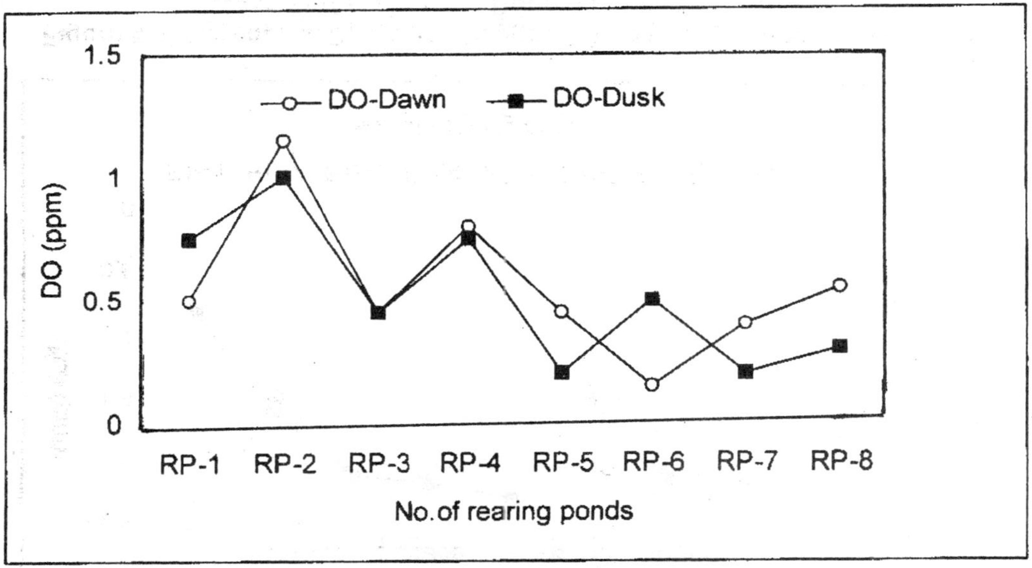

Figure 5: Difference in Dissolved Oxygen Content (ppm) of Water of Ponds during Dawn and Dusk Under Different Pond Management Practices

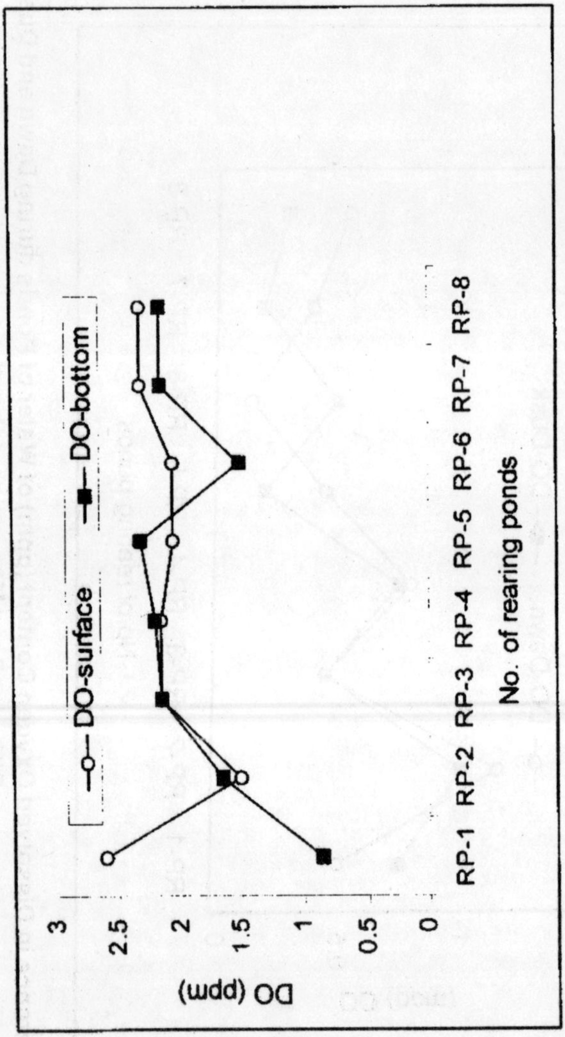

Figure 6: Difference in Dissolved Oxygen Content (ppm) of Water of Ponds at the Surface and the Bottom of Ponds at Dawn

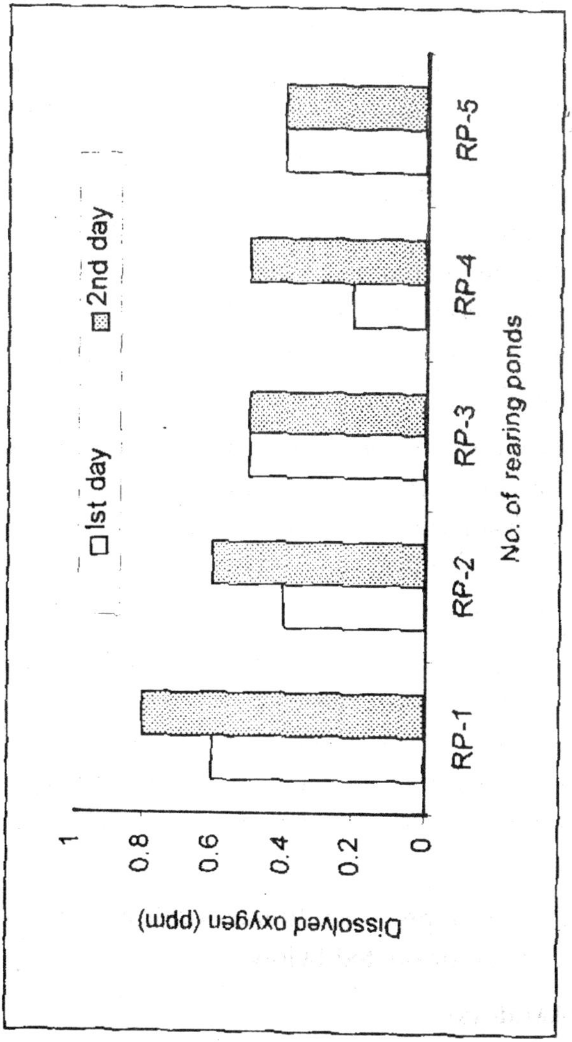

Figure 7: Increase in Dissolved Oxygen Content of Water (ppm) of Different Ponds Due to Aeration in Two Consecutive Days

aerators to be used depend on the level of DO content and the total volume of water.

The solubility of oxygen in to water medium influenced by temperature, pressure and the content of dissolved salts. So an increase in D.O. content of water is never possible without alteration of the factors regulating the solubility of oxygen into the medium of water. But an increase in D.O. content of water with aeration is caused due to exposure of greater surface area due to aeration. Moreover, the dissolved oxygen content of water always remain lower than the theoretical value even under normal pond environmental conditions, which cause an increase in D.O. due to aeration.

Aerators are mechanical devices that increase the rate at which oxygen enters the water. The techniques of aeration are (i) splashing of water into the air or (ii) released of air bubbles into the water are known as splashers and bubble aerators.

The oxygen transfer efficiency of aerators is expressed as kg/O_2 transferred per kilowatt hour of power applied to aerators. The cost of transfer oxygen to pond water with paddle wheel aerators is lower compared to other aerators. However, all type of aerators could be used in aquaculture with success.

The different type of aerators used for aeration of fish ponds have been presented below.

Type of Aerators

Paddle wheel

Propeller-aspirators pumps

Vertical pumps

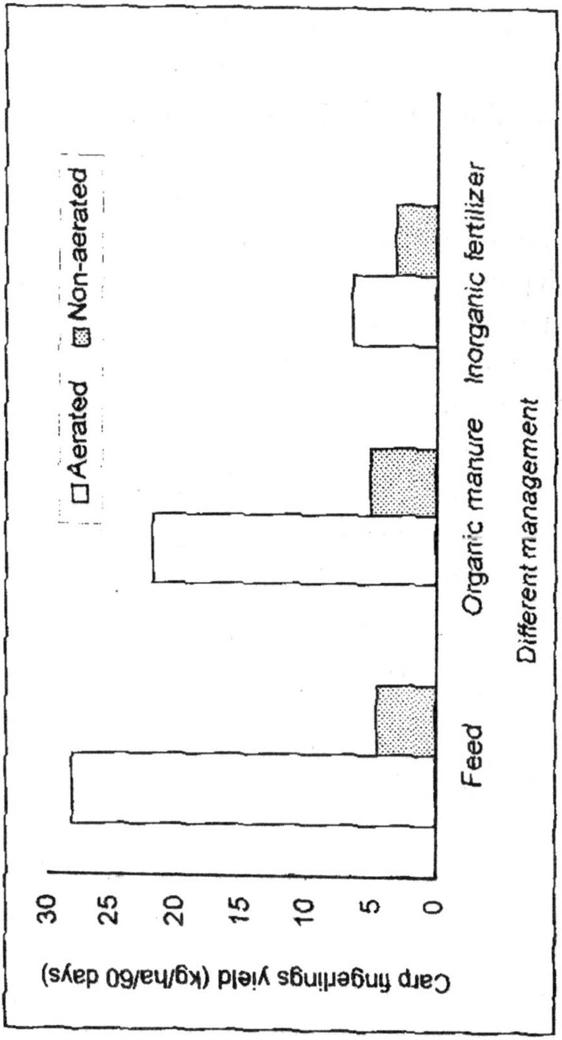

Figure 8: Effect of Aeration on Production of Carp Fingerlings Under Different Pond Management Practices

Pump sprayer

Diffused air system

10.8 Water Circulation

Circulation of water also serve the purpose similar to aeration in fish ponds. It prevents thermal and chemical stratification in ponds and thereby makes the entire pond more habitable by fish and eliminates oxygen depletion at the mud water interface. Aeration also cause circulation of water in fish ponds and this considered to be more efficient than other paddle wheel aerators for water circulation in ponds than other types in circulation of pond water. In absence of aerators water pumps are also used for circulation of water in ponds.

Besides improvement of dissolved oxygen content of water, aeration or circulation of water also enhance carp yield, degradation of soil organic matter and better availability of nutrients under normal and different pond management practices (Figure 8).

Moreover, a marginal difference in D.O. content between the surface and bottom of nurseries prior to stocking with carp spawn is desirable to achieve higher survival of carp fry in short term rearing in nurseries.

Chapter 11
Heavy Metals

Presence of high concentrations of heavy metals have been reported in polluted waters in many countries. Thus, there is an interest in the toxicity of heavy metals to fish and other aquacultural species. Information on the toxicity of heavy metals to a variety of species of freshwater Indian major carps are still lacking. However the safe level of some of the heavy metals recommended by US Environmental Protection Agency have been presented in Table 12.

The safe levels recommended by U.S. Environmental Protection Agency are conservative estimates, which are 10 to 100 times lower than the lowest concentrations which have been reported to harm organisms in laboratory toxicity tests.

In ponds, heavy metals are adsorbed on to suspended clay particles and chelated with the organic matter present in water. Some of the heavy metals also form complexes with oxides, hydroxides and carbonates in water. The toxicity is related primarily to the dissolved, ionic form of the metal, *e.g.*, Cu^{2+} rather than to adsorbed, chelated or complex forms, thus only a small percentage of the heavy metals remain in ionic form in most waters.

Table 12

Metal	*Safe Level Recommended by US Environmental Protection Agency (Mg/l)*
Cadmium	10
Chromium	100
Copper	25
Lead	100
Mercury	0.1
Zinc	100

Chapter 12
Pesticides

Pesticides used for agricultural crops sometimes find their way into fish pond as runoff waters and cause mortality of fishes. Chlorinated hydrocarbon insecticides have the greatest potential for harming fish. The safe level of some selected insecticides recommended by US Environmental Agency to a wide array of freshwater and marine animals have been presented in Table 13.

Apparently, the use of pesticides with long residual lives is declining in most countries and many pesticides that are used today degrade to non-toxic forms within a few days. However, pesticides are potentially harmful until they are degraded. The use of pesticides in fish farming area should be discouraged.

Table 13

Pesticides	*Safe Level Recommended by US Environmental Protection Agency (Mg/l)*
Aldrin/Dieldrin	0.003
BHC	4.0
Chordane	0.01
DDT	0.001
Endrin	0.004
Toxaphone	0.005

Chapter 13
Aquaculture and Pond Management

A considerable increase in carp yield have been achieved in recent years with the introduction of various inputs and adoption of improved technologies in culture systems (Figures 9 and 11). But increase is restricted beyond a certain limit in presence of several limitations and its impact on pond environment is essential for aquaculturists (Figure 10).

Aquaculture is a three tier system comprising of rearing of carp spawn to fry, fry to fingerlings and fingerlings to adult fish while the period of rearing vary from the weeks to one year and are conducted/carried out in nurseries, rearing and stocking ponds. Whereas the real culture system is related to the rearing of fingerlings to adult stage.

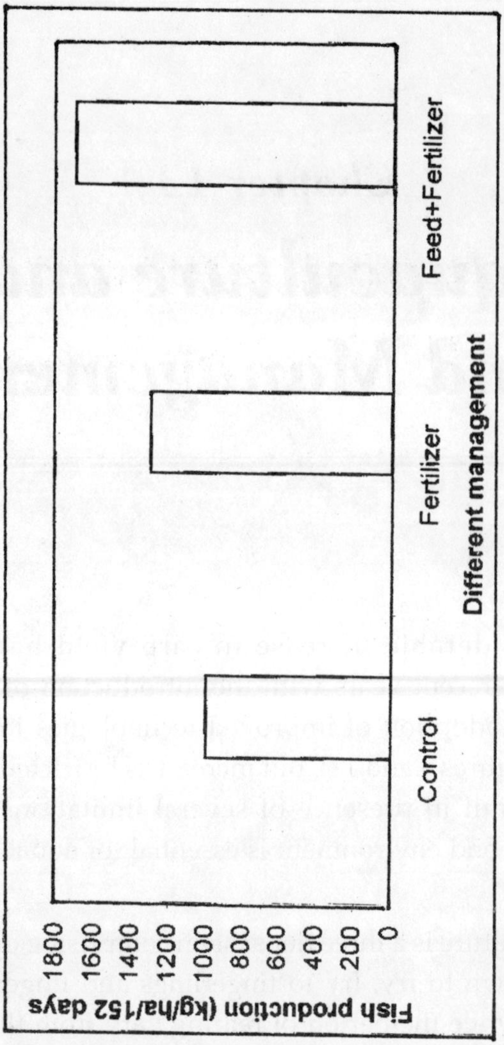

Figure 9: Fish Production Under Different Pond Management Practices

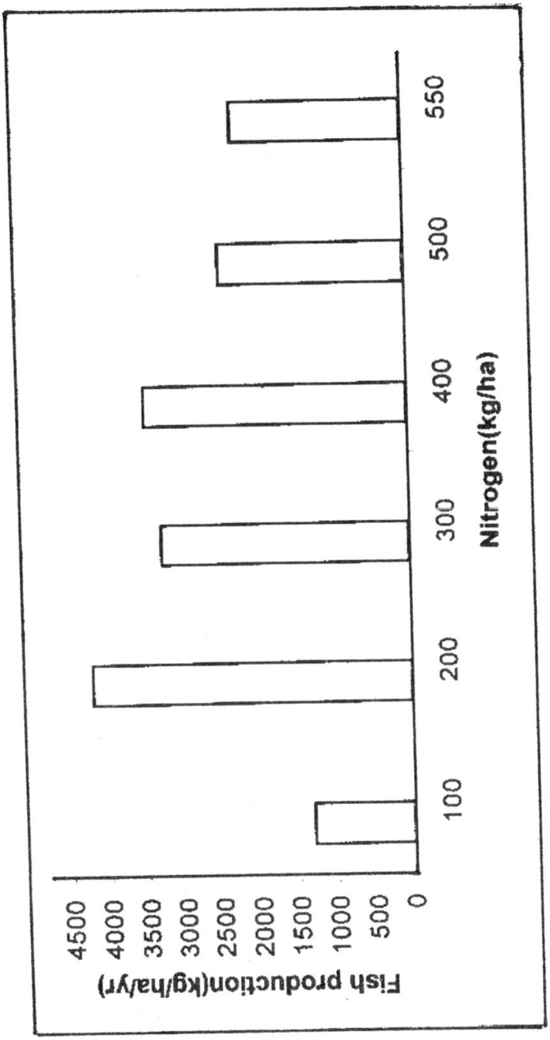

Figure 10: Fish Production at Different Level of Added Nitrogen

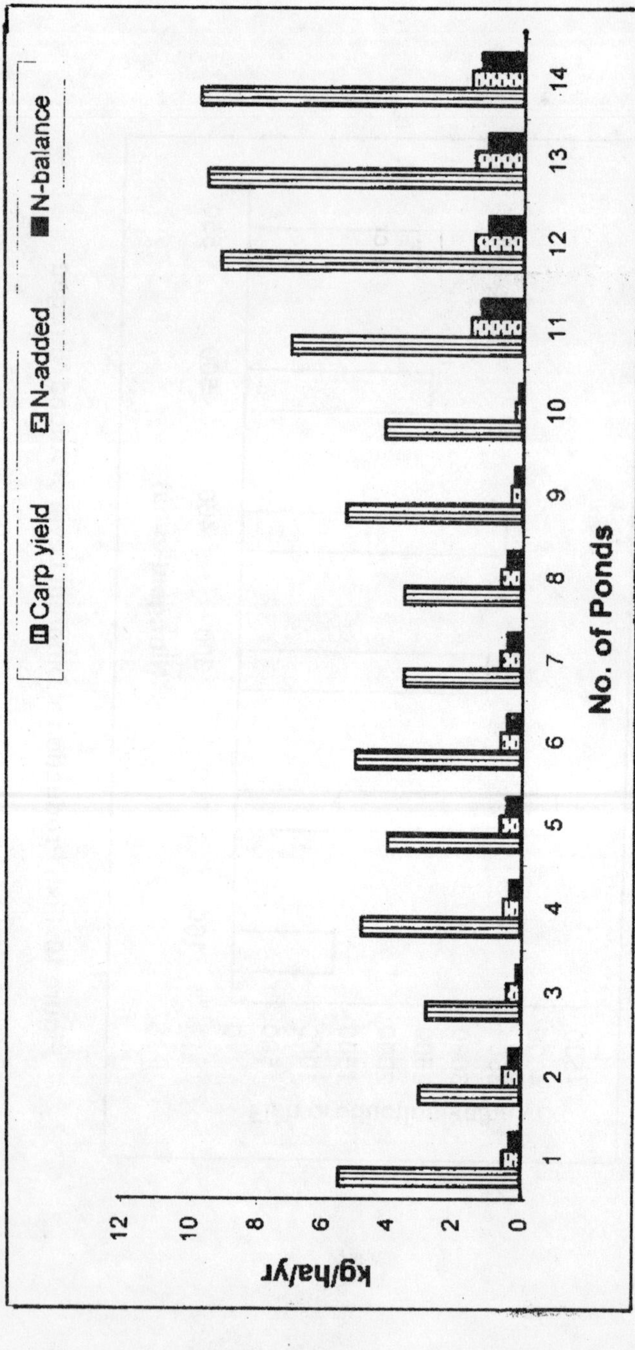

Figure 11: Relationship Between Carp Yield, Nitrogen Added and its Balance

Further the culture practices are broadly divided into three categories into extensive, semi-intensive and intensive or super-intensive on the basis of density of stocking. In extensive system the density of stocking remains mostly 5000 fingerlings/ha while in semi-intensive and intensive system it varies from 5000–7500 and 7500–10,000 or even upto 15,000 fingerlings/ha respectively. While the management measures varied from fertilization, manuring with or without feed and carp yield was comparative higher with the combination of both fertilizers and feed compared to fertilizers and manures only.

As a consequence of variation of the culture practices from extensive to intensive system with the adoption of various management measures though (carp yield has been increased considerably) or a considerable increase in carp yield has been achieved but such unlimited increase in yield in culture operation is restricted beyond a certain limit in presence of some limiting factors.

Chapter 14
Limiting Factors

Biological, chemical and economic are the factors very often restrict the production in culture systems.

14.1 Biological

Impoundments or water bodies of various sizes are very often used for aquaculture and the culture practices vary according to the size and water retention period in rearing of spawn to fry, fry to fingerlings and fingerlings to adult fish. In most cases, fry and fingerling rearing are undertaken in smaller water bodies while large fish culture made in larger water bodies.

However, lack of information is available so far on the ideal size of ponds for various culture practices. 0.1 to 1.0 ha ponds are normally used for culture of adult fishes and are considered suitable for adoption of efficient management measures though 1.0 ha ponds have been utilized for rearing of carp spawn recently.

Besides size of pond, depth of water in pond or the volume of water available during the culture period is also an important aspect for different culture practices. The average depth of water usually fluctuates from 1.0 to 1.5 meters in summer to 2.5 meters in post monsoon period in most aquaculture ponds while more than 2.5 meters may cause both chemical and thermal stratification in ponds. Thus, maintenance of desired water depth during the entire growing period in production ponds is also important to achieve higher yield but is a major constraint in aquaculture systems. In some cases, ground water or bore well waters are also used to maintain the water level in ponds. The suitability of such waters for aquaculture also have an impact on culture system.

Increase in the density of stocking from 5000 to 10,000 to 15,000 fingerlings per hectare with a combination of different species for better utilization of available food in the environment and introduction of various inputs it has been possible to achieve higher yield. Even higher yield has been reported at a stocking density of 20,000 fingerlings per hectare under composite carp culture. But increase in the density of stocking arbitrarily in a limited space and volume of water without considering the growth potential of individual fish under normal conditions and altered conditions within a fixed culture period will obviously affect the growth of individual fish due to limited space compared to low density of stocking with available space for growth. Moreover, in spite of similar yield under both low and higher density of stocking a considerable variation in average growth of individual fish has been reported under productivity and carrying capacity of pond.

Moreover, lack of information on the water requirement for better growth of an individual adult fish and also in absence of concensus on the harvestable size of an individual fish are also other important factors in deciding the density of stocking. However, assuming that the water requirement of an individual fish being $1m^3$ and the harvestable size of 1.0 kg of an individual fish after a growing period of one year, the stocking density becomes 10,000 fingerlings/ ha having an average depth of water of one meter.

Even, at a stocking density of 10,540 fingerlings/ha with a combination of five species and provided with fed and fertilizers, it has been reported that the growth of silver carp, common carp, catla, rohu and mrigal after one year were 1032-1152,1516-1772, 1179-1197, 751-854 and 657-766 gm. respectively while the corresponding figures were 1094-1149, 408-422, 562-570, 500-563 and 364-621 gm with fertilizers only at a stocking density of 7500 fingerlings per hectare. Thus, it is obvious that even at stocking density from 7500-10,000 fingerlings per hectare failed to produce an uniform growth of one kilogram of an individual carp of each species after one year under composite carp culture.

Further, as a consequence of increase in the density of stocking it is obvious that larger amount of inputs in terms of feed and fertilizers has to be introduced into the system for enhanced production. While introduction of such an amount of inputs ultimately have adverse effect on pond environmental conditions in terms of deficiency of oxygen, accumulation of metabolites besides growth of aquatic animals.

14.2 Chemical

Soil and water are the integral component of pond environment and a dynamic equilibrium is always maintained in environmental conditions due to existence of an interrelationship between the two phases. Interference of the equilibrium in existence with the introduction of various inputs as feed, fertilizers and manures will affect the existing environmental conditions, ultimately effect growth and yield.

As a consequence of such interference on the existing equilibrium will lead to consequent removal or conversion of nutrients added as fertilizers, manures and feed so as to restore the equilibrium. Thus, pond having a fixed area and volume of water is able to accept a limited amount of nutrient loading without saturating or polluting the water. Further, addition of nutrients through introduction of inputs beyond this limit will either make the water saturated with nutrients or pollute the water. But such condition never occur due to the existence of the equilibrium between the two phases. Moreover, the existence of an equilibrium between the two phase prevent such condition to occur in pond environment. Under such condition, higher the nutrients added to the system, the faster is their removal from the water phase. Moreover, increase in primary production always does not increase in carp yield in presence of other limiting factors and thereby restrict the amount of added nutrients. According to reports, concentration of dissolved inorganic nitrogen and phosphorus of productive waters remain in the range of

0.5-1.0 ppm and 0.2-0.5 ppm respectively while in polluted waters the concentration of dissolved inorganic N remain higher than 1.0 ppm.

Thus, considering the optimum concentration of dissolved inorganic N and phosphorus of productive water being minimum of 1.0 and 0.5 ppm respectively with a N:P ratio of 2:1 or 1.0 and 0.25 ppm with a ratio of 4:1, the amount of nitrogen and phosphorus required to raise their concentration to the desired level is 240 and 120 kg/ha/yr respectively applied in twenty four equal installments at an interval of fifteen days for one hectare pond with 1.0 m average depth of water. Accordingly, optimum level of

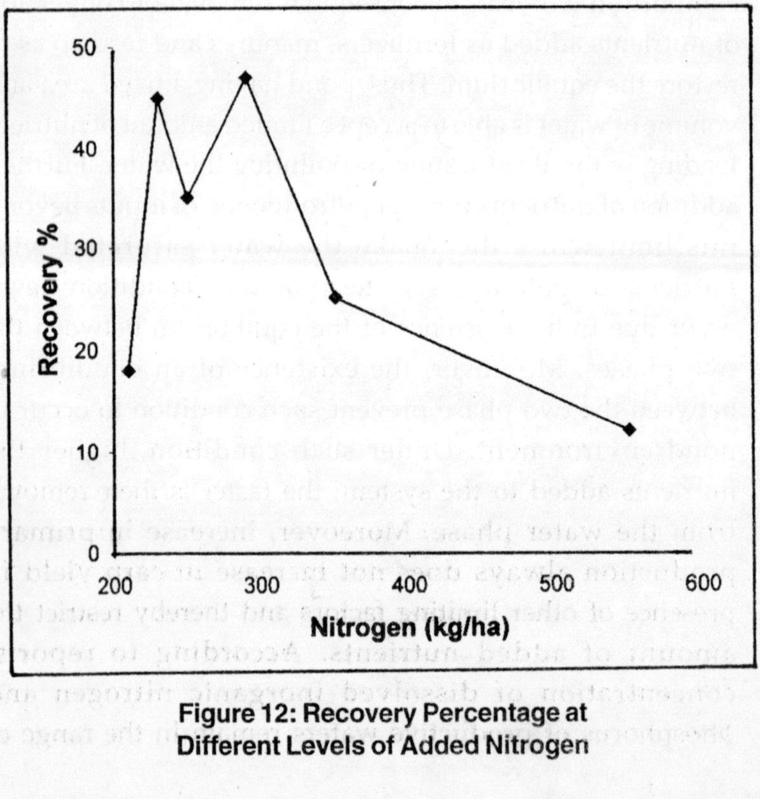

Figure 12: Recovery Percentage at Different Levels of Added Nitrogen

nitrogen and phosphorus recommended for application to ponds is 250-300 kg/ha/yr and 125 to150 kg phosphorus /ha/yr.

14.3 Economic

Besides biological and chemical limitation economic return of added nutrients also become a limiting factor in aquaculture. The increase in yield tend to increase with the corresponding increase in added nutrient beyond which the yield does not correspond proportionately with simultaneous increase in added nutrients. Such a level of yield beyond which further addition of inputs does not result in economic return of the added nutrients corresponds to economic limitations (Figure 10). Some of the results of composite culture indicate that while the increase in carp yield was 3.5 times (2275 to 7719 kg/ha/yr) the simultaneous increase in added nitrogen was nearby six times (210 to 1207 kg/ha/yr) (Figure 12).

Chapter 15
Efficiency of Nutrient Utilization

The efficiency of utilization of nutrients added in terms of fertilizer, manures and feed in aquaculture is considered in terms of their conversion into protein, fat and carbohydrate of the reared animals. Accordingly, the efficiency reflects the removal of the added nutrients in terms of ultimate production as fish. Some workers have observed that the efficiency is comparatively higher with nutritionally balanced feed provided in intensive aquaculture compared to extensive system with the application of inorganic and organic either singly or in combination rather than feed. Accordingly, in intensive culture provided with feed the efficiency of utilization in terms of removal of nitrogen and phosphorus as fish were from 21 to 33 per cent and 11 to

28 per cent respectively compared to 5 to 25 per cent Nitrogen and 5 to 11 per cent phosphorus in semi intensive culture systems provided with fertilizers, manures with limited use of feed or without feed.

Similar observations also have been reported on the recovery of carbon (65 per cent), Nitrogen (31 per cent) and phosphorus (32 per cent) as fish in intensive culture systems with intensive feeding. The recovery of nitrogen as fish in static water composite carp culture varied from 5.5 to 30.7 per cent while in extensive to intensive polyculture and provided with feed, fertilizers manures and aquatic weed either singly or in combination, the recovery varied from 14.1 to 47.3 per cent of the added nitrogen. The maximum recovery from 43.6 to 47.3 per cent was observed in the ratio of 2:1 of N:P with the application of fertilizers and manures in culture practices (Figure 13). Further, according to reports, the percentage of removal of total nutrients introduced into the system were of smaller magnitude under both semi intensive (15 per cent N and 8 per cent P) and intensive (N-27 per cent and 15 per cent P) culture system and the difference on the removal of nutrients from the two systems were attributed to higher nutrient conversion efficiency of N (73 per cent) and P (88 per cent) in intensive culture system.

A comparatively lower conversion efficiency under semi-intensive than intensive system is due to involvement of one additional step in the conversion of added nutrients in terms of fertilizers and manures for production of natural fish food organisms. Some workers have also been reported similar observations on the efficiency of conversion of nutrients added as pig and poultry manures were much

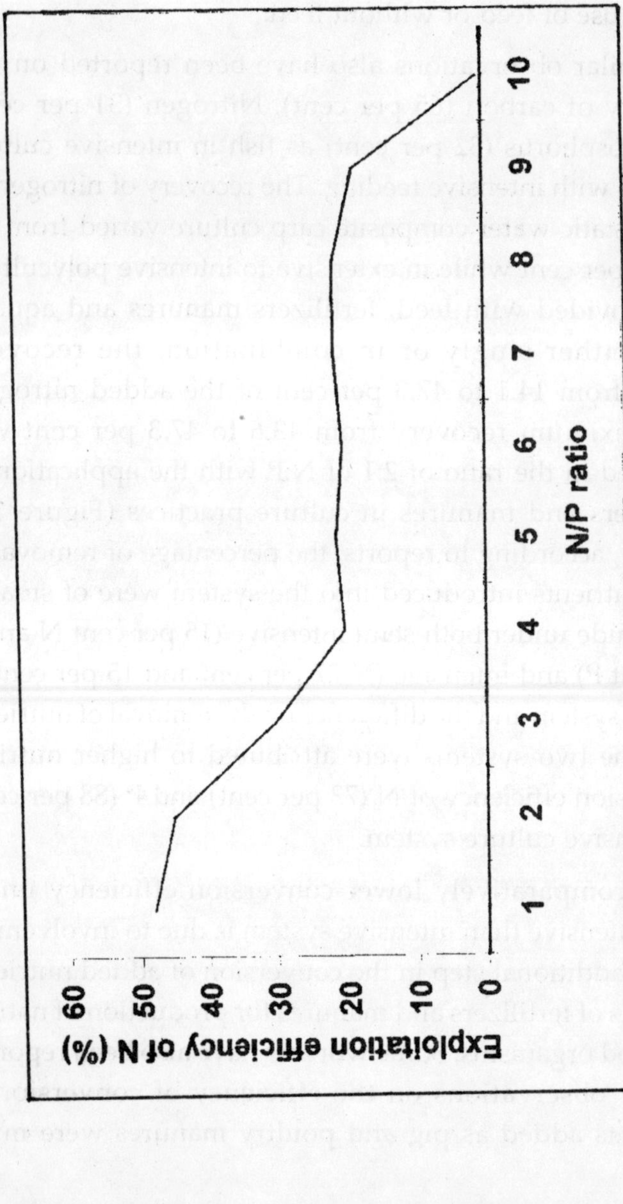

Figure 13: Relationship Between Exploitation of Nitrogen and N/P Ratio

less in semi-intensive culture system than that of intensive system. Further, at higher density of stocking and intensively fed intensive culture system only 65 per cent carbon, 11 per cent N and 32 per cent P were removed as fish. Such observations indicate the importance of carbon, in culture system besides nitrogen and phosphorus and its removal is comparatively more than that of nitrogen and phosphorus. The exploitation efficiency of nitrogen added in static water fish ponds was maximum (43.6 to 47.3 per cent) with the application of 200 to 300 kg/N in combination with phosphorus as inorganic fertilizers appeared to be the optimum level of nitrogen which could be applied for increased production. Application of such an amount of nitrogen at twenty four equal fortnightly intervals to one hectare pond with an average depth of water of 2.0 and 1.0m will theoretically raise the dissolved inorganic nitrogen concentration to 0.625 and 1.25 ppm respectively which is considered to be the desired level of this element of productive waters. Introduction of a common feed for both Indian major carps and exotic carps having diverse food and feeding habits, will obviously affect efficiency of utilization of nitrogen in composite carp culture.

Chapter 16
Impact of Aquaculture on Pond Environment

Informations on the impact of culture system on pond environment has drawn the attention of aquaculturists in recent years is of great significance for aquaculture. As a consequence of variations in culture systems, species and management practices in aquaculture, detailed information on their impact on pond environment in terms of nutrient balance, residual effect, variation in yield in successive years in relation to nutrient build up are lacking. However, the limited information available on these aspects have been presented in this chapter as a guideline for aquaculturists.

It is evident that a considerable increase in carp yield is feasible with the variation in culture system from extensive to intensive system and management practices. In spite of increased yield a major part of the nutrients added are lost in the environment and their magnitude varies depending on the variation in culture and management practices. Further, low recovery of nutrients also signifies that a major part of the nutrient are either lost or sedimented at bottom sediment, while a part remain only in the water phase except in case of intensive culture practices with water exchange system compared to semi intensive system with annual draining of water. The observation of some workers on tentative nutrient balance in intensively fed static water ponds were similar to those reported for semi-intensive culture system fertilized with manure and feed of live stock.

The dissolved N and P content of water does not increase in time, but varied about a mean value of 1.5 and 0.5 mg/1 respectively. Under such conditions 75 per cent nitrogen and 50 per cent phosphorus provided in the feed remain unutilized by fish and most of the nitrogen (1.0 ton/ha) accumulated in the bottom sediment. While the concentration of dissolved inorganic nitrogen remained 0.192 to 0.969 ppm after the experiment. A similar absence of increase in nutrient concentration in water also have been reported in culture of Nile Tilapia in earthen ponds with the use of urea and triple superphosphate supplemented with chicken manure. The interdependence of soil and water restore the equilibrium in existence in pond environment between the two phases with consequent removal of the added nutrients from the water phase. But introduction of large amount of either feed or organic manures into the system will cause the presence of higher level of dissolved inorganic N and P in water and as a result either depletion of oxygen or undesirable aquatic blooms will be developed. According to reports, the magnitude of removal of N and P under such system were 15.1 and 8.1 per cent respectively and the total N and P concentration remained 1.12 and 0.91 mg/l respectively. The presence of such concentration of N and P in water is considered to be of much higher than the desired level of these nutrients of productive waters and accounted only 1.8 per cent and 5.9 per cent in water while the balance disappeared in the system.

The removal of such magnitude of N and P does not pollute the water body under semi-intensive system as most of the nutrients added are immobilized in the bottom sediment in contrast to intensive culture system, where 73 per cent N and 86 per cent P pass through the system pollute the external environment. In absence of water exchange in intensive culture system, it is obvious that a considerable amount of nutrients added remain immobilized in the bottom sediment due to interrelation between water and sediment. But in practice a major part of the added nutrients remain both in water and sediment in intensive culture systems, in spite of presence of such high magnitude of nutrients in the sediment remain unavailable for production processes. Accumulation of such high amount of nitrogen in the sediment will enhance the organic matter content of the bottom sediment due to immobilization of nitrogen in organic combination. Increase in organic matter content in time is a natural process in water lodged or submerged soils similar to pond sediment and such increase is comparatively higher in fertilised ponds compared to unfertilized ponds. A considerable increase in the carbon and nitrogen ratio of the bottom sediment from an initial value of 3.6-17.3 to 4.9-29.0 has been reported for composite carp culture under different management practices. Besides, carbon nitrogen ratio in presence of high organic content in bottom sediment will also affect pond environment adversely such as increase in soil acidity, reduced bacterial activity affecting recycling of nutrients and ultimately make the pond unproductive/ lead to unproductivity. On the basis of comparison made on the polluting capacity of semi-intensive systems (combination of inorganic and organics) and intensive systems (supplementary feed having FCR 1.0-2.5 with or without fertilizers), it has been reported that intensive systems are more polluting (26-44 times) than semi-intensive systems (12-15 times) expressed as a percentage of nitrogen required for 1.0 kg. fish. However on the basis of weight of nutrients released to the environments the difference is reduced in intensive system due to high conversion efficiency of nitrogen and phosphorus than semi-intensive system. While in terms of nitrogen and phosphorus released to the environment, intensive systems are more polluting (7-31 times) than semi-intensive (3-11 times) system.

Thus from the informations so far available on the impact of intensive and semi- intensive culture systems on pond environment, it is evident that semi-intensive culture systems with balanced use of nutrients (carbon, nitrogen and phosphorus) in suitable form and amount are more desirable in reducing the adverse effect on nutrient loading in pond environment and in terms of both efficient utilization of added nutrients and sustainable yield in aquaculture.

Bibliography

Avnimelech, Y. and Lacher, M., 1979. A tenative nutrient balance for intensive fish ponds. *Bamidgeh,* 31(1):3-8.

Avnimelech, Y. and Lacher, M., 1984. A method for evaluation of conditions in a fish pond sediment *Aquaculture* 23:361-365.

Avnimelech, Y., Diab, S. and Kochava, M., 1992. Control and utilization of inorganic N in intensive culture ponds. *Aquaculture and Fisheries Management,* 23:321-430.

Banerjee, S. M., 1967. Water quality and soil condition of some states of India in relation to fish culture *Ind. J. Fish* Vol.11(1 & 2):115-144.

Banerjee, S. M. and Ghosh, A. N., 1963. Soil nutrients and plankton production in fish ponds. *Ind. J. Fish* Vol.10(2A):627-633.

Banerjee, S. M. and Mondal, L. N., 1965. Inorganic transformation of water soluble phosphates added to fish ponds as influenced by the nature of pond soils. *J. Inland Fish Soc.* India, 13:167-173.

Banerjee, S. M. and Banerjee, S. C., 1967. Fertilization of fish ponds with trace element manganese for increased production of plankton. *FAO. Fish Rep.* 44 Vol. 3:132-152.

Banerjee, S. M. and Ghosh, S. R., 1970. Studies on correlation between soil reaction and different bound phosphorus in pond soils. *J. Inland. Fish. Soc. India,* 2:113-120.

Bear, F.E. (ed.) 1964. Chemistry of the soil. Oxford and IBH Publishing Company, New Delhi, 515 p.

Boyd, C. E., 1979. Water quality in warm water fish ponds. Alburn University AL., USA. 359 pp.

Boyd, C. E., 1970. Influence of organic matter on some characteristics of aquatic soils. *Hydrobiologia* 36:17-21.

Boyd, C. E., 1982. Water quality management for pond fish culture. Elsevier Publishing Company, Amsterdam, Oxford, New York, 318 P.

Boyd, C. E., 1976. Chemical and Textural changes of muds from different depth in ponds. *Hydrobiologia.* Vol.48(2):141-144.

Boyd, C. E. and Cuence Michel, L., 1980. Refinements of the lime requirement procedure for fish ponds. *Aquaculture,* 21:293-299.

Boyd, Claude, E. and Musing Yont, 1981. Orthophosphate uptake by plankton and sediment. *Aquaculture,* 22:165-173.

Boyd, Claude, E., Musing Yont and Tucker Luther, 1981. Effects of three phosphorus fertilizers on phosphorus concentration and phyto plankton production. *Aquaculture,* 22:175-180.

Bouldin, D. R., Johnson, R.L., Burda, C. and Kao, Chun-Wei, 1974. Losses of inorganic N from aquatic systems. *J. Environ Quality,* 3:107-114.

Buckman, Q., Hary and Brady Nyle, C. *The nature and properties of soils.* Eursia publishing House Pvt. Ltd., New Delhi, 567 P.

Chatterjee, D. K., 1986. Organic and Inorganic Fertilizers in pond manuring- International Training Programme on Composite carp culture, CIFA, Kausalyaganga Sept'01 to Nov'15, 1986.

Chatterjee, D. K., 1987. Water quality for carps and catfish ponds. USAID Training Programme on Freshwater Aquaculture, CIFA, Kausalyaganga, July, 14–Aug, 13, 1987.

Chatterjee, D. K. and Saha, G. N., 1986. Effect of lime on fish pond soils. *J. Inland Fish Soc. India,* 18(1):41-44.

Chatterjee, D. K., 1997. Exploitation Efficiency of added Nitrogen and its effect on pond environment. *J. Aqua. Trop.* 12:124-131.

Chakraborty, R. D., Sen, P. R. and Chatterjee, D. K., 1975. On the use of fertilizers and supplementary feed for enhancing fish production in freshwater ponds. *J. Nat. Acad. of Sci. Ind.* XLV (III) Sec B:192-196.

Chakraborty, R. D., Sen, P. R., Chatterjee, D. K. and Kowtal, G. V, 1973. Observations on the relative usefulness of different feed for carp spawn and fry. *J. Inland. Fish Soc., India,* Vol.V, 182:183.

Chatterjee, D. K., 1989. Role of soil and water in Aquaculture. Summer Institute of Fish Diseases Diagostate and Health management in Freshwater Aquaculture system, 5-24th June, 1989.

Chatterjee, D. K. and Saha, P. K., 1996. Characteristics of organic matter in undrainable fish pond soils. *The Isralei Journal of Aquaculture,* Bamidgeh, 48(3):142-148.

Chatterjee, D. K. and Saha, P. K., 2000. Effect of organic manures on organic carbon contents of submerged pond soil. *The Israeli Journal of Aquaculture,* Bamidgeh, 52(3):1118-126.

Chaudhuri, H. L., 1974. Record Fish Production with Intensive culture of Indian and Exotic carps. Current Sci. Vol.XLIII No.10:303-304.

Chen, T. P., Lin, 1967. Increase of production in freshwater fish ponds by the use of inorganic fertilizers. FAO Fish Rep., 44(3):210-225.

Das, B. C., 1967. Effect of micro-nutrients on the survival and growth of Indian carp fry. FAO. Fish Rep. 44(3):241-256.

Edwards, P. 1993. Environmental issues in Integrated Agriculture, Aquaculture and waste water Fed Fish culture system, PP.139-170 in RSV Pulin, H. Rosenthal, and J. L. Mclean (Eds). Environment and Agriculture in Developing Countries. ICLARM. Conf. Proc., 31-359P.

Hepher, B., 1965. The effect of impoundment on chemical and Textural changes in Fish pond bottom soils. *Bamidgeh*. Vol.17(3):71-80.

Hepher, B., 1967. Some limiting factors affecting the dose of fertilizers added to fish ponds FAO Fish Rep. 44(3): 1-6.

Matida, Y., 1967. The role of soil in Fish pond productivity in Asia and the Far East FAO Fish Rep. 44(3): 59-63.

Mohanty, A. N., Chatterjee, D. K., Saha, P. K. and Pani, K. C., 1994. Effect of varying C/N ratio on the mineralization of organic nitrogen in fish pond soils. *J. Aqua Trop*. 9:9-14.

Mondal, L. N., 1978. Physico-chemical properties of submerged soils in relation to fertility-Summer Institute on chemistry of submerged soils at BCKKV Kalyani, India.

Patrick, W. H. Jr. and Gotoh, S., 1974. The role of oxygen in Nitrogen loss from flooded soils. *Soil Science*. Vol.118 (2):78-81.

Patrick, W. H. Jr. and Tusneem, E., 1972. Nitrogen loss from flooded soils. *Ecology*, Vol. 53: 735-737.

Prowse, G. A., 1967. A review of the methods of fertilizing warm water fish ponds in Asia and Far East. FAO Fish Report, 44(3):7-12.

Rabanal, R. H., 1967. Inorganic fertilizers for pond fish culture. *FAO Fish Rep*. 44(3): 164-178.

Raman, K., Gosh, S. R. and Chatterjee, D. K., 1975. Studies on the ecology of fish ponds with and without fish with special reference to Bottom Fauna. *J. Inland Fish Soc. India* Vol. III, 173-181.

Ramaprabhu, T., Tripathi, S.D., Chatterjee, D.K., Jena, S. and Das, K.M., 1990. Use of ammonia as a fish toxicant in pond Ecosystem *Aquaculture Hungarica*. Vol.VI: 97-104.

Saha, G. N., 1969. A note on the effect of water level on the availability of nitrogen in pond soils. *J. Inland Fish Soc. of India,* 7:135-136.

Saha, G. N., 1979. Techniques of pond fertilization and use of fertilizers in freshwater Aquaculture for increased Fish Production. *Fert. News* 24(2):3-6.

Satomi, Y., 1967. Physiological significance of carbon sources in fertilized fish ponds. FAO Fish Rep.44(3):257-264.

Saha, G. N. and Chatterjee, D. K., 1977. A preliminary study on the comparative efficiency of two mixed fertilizers in both productive and unproductive soils. *J. Nat. Acad. Sci., India*, 47(5) II : 93-100.

Saha, G. N. and Chatterjee, D. K., 1975. Urea for enhancing the production of Zooplankton in fish ponds. *Sci and Cult*. 41(7):320-321.

Saha, G. N., Chatterjee, D. K. and Raman, K., 1974. Observations on manuring nursery ponds with only chemical nitrogenous fertilizers for successful rearing of carp spawn. *J. Inland. Fish Soc. of India*. Vol.VI:162-166.

Saha, G. N., Chatterjee, D. K., Selvaraj, C. and Mazumdar, N. N., 1978. A record of increased fish production in freshwater ponds by use of Chemical fertilizers alone *Sci & Cult*. Vol.44:423-424.

Saha, P. K., Chatterjee, D. K. and Tripathi, S. D., 1989. Composition and Nature of humus in some Indian Fish Pond Soils. *J. Aqua. Trop.* 4:137-142.

Saha, G. N., Chatterjee, D. K., Selvaraj, C. and Mazumdar, N. N., 1979. Role of some trace elements pond in fertilization (Abs) Proc. *66th Ind. Sci Cong. Fert*-III.48.

Saha, G. N., Chatterjee, D. K., Selvaraj, C. and Mazumdar, N. N., 1979. Evalution of chemical fertilizers in enhancing fish production (Abs). *Symp. Inland Aquaculture,* Barrackpore, Feb, 12–14th Feb., 48 P.

Saha, G. N., Raman, K., Chatterjee, D. K. and Gosh, S. R., 1975. Relative response of three Nitrogenous fertilizers in different pond soils in relation to primary productivity, plankton and survival and growth of *Labeo rohita* spawn. *J. Inland Fish Soc. India* Vol.VII:162-172.

Saha, P. K., Adhikari, S. and Chatterjee, D. K., 1996. Available iron, copper, zinc and manganese in some freshwater pond soils of Orissa in relation to soil characteristics. *Ind. Soc. Soil Sci.* Vol.44(4):681-684.

Sen, P. R. and Chatterjee, D. K., 1976. Enhancing production of Indian major carp fry and fingerlings by use of growth promoting substances. *FAO. Tech. Conf. on Aquaculture* Kyoto, Japan. 26th May-2 June, 1976.

Sen, P. R. and Chatterjee, D. K., 1978. Increased production of major Indian carp fry by addition of growth promoting substances-*Symp. on Fin Fish Nutrition* and *Feed Technology*. Hamburg, 20-27th June, 1978.

Sinha, V. R. P. and Ramachandran, V., 1985. Freshwater Fish culture Pubn. Information Divn. *ICAR*, New Delhi, 80 P.

Sinha, V. R. P., Khan, H.A., Chakraborty, D.P. and Gupta, M.V., 1980. Preliminary observation on nitrogen balance of some ponds under composite fish culture. *Aquaculture Hungarica.* Vol.II:105-116.

Swingle, H. S. and Smith, E. V., 1938. Fertilizers for increasing the natural food for fish in ponds. *Trans Amer. Fish. Soc.*, Vol.68:126-135 P.

Szumiec, Maria, 1967. Thermal characteristics of carp ponds *FAO Fish Rep.* 44(3):265-273.

Tal, S., 1962. An exposition of the terms "Pond Productivity and Carrying capacity of ponds. Bamidgeh. Vol.II No.3:49-51.

Wiebe, A. H., 1929. The effects of various fertilizers on plankton production. *Trans Amer Fish Soc.* Vol. 59: 94-105.

Wrobel, Stanley, 1967. The role of soil in fish production in ponds. *FAO Fish Rep.* 44(3):153-163.

Yawalkar, K.S., Agarwal, J.P. and Bokde, S., 1977. Manures and Fertilizers Agri. Horticultural Publishing House. Nagpur, 300 P.

Zur, O., 1981. Primary production in intensive fish ponds and a complete organic carbon balance in the ponds. *Aquaculture*, 23: 197-210.

Index